Blast effects on buildings

Design of buildings to optimize resistance to blast loading

Blast effects on buildings

Design of buildings to optimize resistance to blast loading

Edited by
G. C. Mays and P. D. Smith

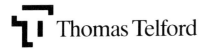
Thomas Telford

A Structural Design Guide prepared by a Working Party convened by the Structures and Buildings Board of the Institution of Civil Engineers.

Working Party Membership:

Brigadier C. L. Elliott	Ministry of Defence
P. Jackson	R. T. James and Partners Ltd
Major R. J. Jenkinson	Ministry of Defence
D. Mairs	Whitby and Bird
Prof. G. C. Mays	Cranfield University
P. D. Smith	Cranfield University
I. Thirlwall	R. T. James and Partners Ltd
C. J. R. Veale	Ministry of Defence

Published by Thomas Telford Publications, Thomas Telford Services Ltd, 1 Heron Quay, London E14 4JD

First published 1995

Distributors for Thomas Telford books are
USA: American Society of Civil Engineers, Publications Sales Department, 345 East 47th Street, New York, NY 10017-2398
Japan: Maruzen Co. Ltd, Book Department, 3–10 Nihonbashi 2-chome, Chuo-ku, Tokyo 103
Australia: DA Books and Journals, 11 Station Street, Mitcham 3132, Victoria

A catalogue record for this book is available from the British Library

ISBN: 0 7277 2030 9

© The Authors, 1995, unless otherwise stated.

All rights, including translation, reserved. Except for fair copying, no part of this publication may be reproduced, stored in a retrieval system or transmitted in any form or by any means, electronic, mechanical, photocopying or otherwise, without the prior written permission of the Publisher: Books, Publications Division, Thomas Telford Services Ltd, Thomas Telford House, 1 Heron Quay, London E14 4JD.

The book is published on the understanding that the authors are solely responsible for the statements made and opinions expressed in it and that its publication does not necessarily imply that such statements and/or opinions are or reflect the views or opinions of the publishers.

Typeset by MHL Typesetting Ltd, Coventry

Printed in Great Britain by Redwood Books, Trowbridge, Wilts

Contents

Editors' note	vii
1. Introduction	**1**
C. L. Elliott	
Objectives	1
Scope	1
Terrorism	2
Risk	2
The special effects of catastrophic failure	3
Partial safety factors in blast design	3
A design philosophy: planning for protection	5
Reference	7
2. Basic guidelines for enhancing building resilience	**8**
P. Jackson, I. Thirlwall and C. J. R. Veale	
Introduction	8
The requirements of the client	14
Architectural aspects and design features	15
Provision of shelter areas	16
Location of key equipment and services	19
Building form	19
Design of individual structural elements	20
Building layout	22
References	23
3. Blast loading	**24**
P. D. Smith	
Notation	24
Introduction	25
Explosions	25
Explosion processes terminology	26

Explosives classification	26
Blast waves in air from condensed high explosives	27
Blast wavefront parameters	27
Other important blast wave parameters	29
Blast wave scaling laws	30
Hemispherical surface bursts	32
Blast wave pressure profiles	33
Blast wave interactions	33
Regular and Mach reflection	35
Blast wave external loading on structures	37
Internal blast loading of structures	40
Conclusions	44
References	45

4. Structural response to blast loading   46
P. D. Smith

Notation	46
Introduction	47
Elastic SDOF structure	47
Positive phase duration and natural period	49
Evaluation of the limits of response	50
Pressure–impulse diagrams	52
Energy solutions for specific structural components	57
Lumped mass equivalent SDOF systems	59
Resistance functions for specific structural forms	62
Conclusions	64
References	64

5. Design of elements in reinforced concrete and structural steel   66
G. C. Mays

Notation	66
Objectives	68
Design loads	69
Design strengths	70
Deformation limits	71
Introduction to the behaviour of reinforced concrete and structural steelwork subject to blast loading	72
Introduction to the design of reinforced concrete elements to resist blast loading	77
Introduction to the design of structural steel elements to resist blast loading	89
References	94

6. Implications for building operation   **95**
C. J. R. Veale
    Health and safety regulations   95
    Threat assessment   95
    Pre-event contingency planning   96
    Further considerations   97
    Post-event contingency planning   98
    Reference   98

Appendix A. Simplified design procedure for determining the appropriate level of glazing protection   99

Appendix B. Transformation factors for beams and one-way slabs   102

Appendix C. Maximum deflection and response time for elasto-plastic single degree of freedom systems   110

Appendix D. Design flow chart   119

Editors' note

Wherever possible the notation has been made consistent between chapters. However, there may be some instances where, for clarity, it has been decided to retain the notation used in the source material. For this reason the notation has been separately defined for each chapter. There may also be some instances where the definitions used do not exactly match those used in conventional design standards.

1 Introduction

Objectives
This book is aimed at all engineers and architects involved in the design of building structures, and should enable them to have a better understanding of their own and their client's responsibilities in providing buildings which, in the event of an explosion, minimize damage to people and property.

The intended purpose of this book is to explain the theory of the design of structures to resist blast loading and then to suggest relatively simple techniques for maximizing the potential of a building to provide protection against explosive effects. The book does not encourage designers to produce hardened structures, nor should it be seen as a precursor to a 'super code', although guidance is given as to where to look for further details should they be required. In addition, some broad design objectives concerning the protection of people, the protection of equipment and the minimizing of damage to structures are given. The book is produced in parallel with a companion document sponsored by the Institution of Structural Engineers concerning the structural engineer's response to bomb damage [1].

Scope
Following this introduction putting the subject in context, the book starts with some basic guidelines for enhancing the resilience of buildings to blast loading (Chapter 2). An explanation of the nature of explosions and the mechanism of blast waves in free air is then given. This leads to a discussion of the loading that explosions place on structures, both inside and out, and the concept of an idealized loading is introduced (Chapter 3). The response of buildings, people and equipment to blast loading is illustrated. A method of analysis is suggested whereby the strain energy absorbed is equated to the work done or kinetic energy imparted and the structure reduced to a single degree of freedom (Chapter 4). Methods for the design of structures and individual elements using different

construction materials are then given (Chapter 5). The book concludes by providing a summary of the implications of the foregoing for building operation (Chapter 6).

The scope of the book is limited to blast effects which derive from above ground, non-nuclear, high-order explosions and the design only covers structures that contain a degree of ductility. Only passing mention is made of the effects of fire, secondary fragments and ground shock.

The discussion is not solely concerned with the protection of structures against terrorist explosions, although this remains a strong theme throughout. The authors suggest that, although there is a wide knowledge of the effects of terrorism, many public and corporate buildings and installations have been built and are still being built with little concern for the destructive effects of terrorist attacks. As a result they offer their inhabitants and contents too little protection. It is hoped that this book will provide clues as to how to give reasonable and affordable protection against present, and future, threats. Within this introduction and as a setting to the future chapters, some of the special circumstances of terrorist attacks will be illustrated, the idea of risk assessment introduced and a general philosophy for protection offered.

Terrorism

If modern terrorism is described as the deliberate use of violence to create a sense of shock, fear and outrage in the minds of a target population, several factors in the way we now live have made terrorism much more easy to conduct in modern society.

First, terrorists are able to make use of the media as never before to carry a sense of terror to their target population. Television gives terrorists a political leverage out of all proportion to their other powers. Second, developed societies have become very dependent on complex and vulnerable systems (e.g. railways, airlines, gas pipelines, large shopping areas and business centres) which allows the terrorist many suitable targets. Third, terrorists hide behind the camouflage of normal daily life. This means that almost all effective measures to combat terrorism also carry considerable constraints on individual freedoms, which governments are rightly reluctant to impose, and often will not.

The foregoing leads to several conclusions: terrorism today is much easier to contain than to eliminate; there are few completely acceptable antidotes to it; the prudent design will allow for its effects wherever it is possible and affordable.

Risk

Protection is not an absolute concept and there is a level of protection where the cost of protection provided with respect to the cost of the potential loss is optimized. Protection can never offer a guarantee of safety;

conversely, too much protection is a waste of resources with regard to what is being saved. Furthermore, the consequences of loss vary; some loss is incremental, but certain losses, such as human life, essential records or specialist equipment are catastrophic. For these reasons a risk assessment should be made to assess a combination of the type, the likelihood and the consequences of an attack. Some risks will have to be accepted, while others must be deflected at all costs. Advice on how to conduct a terrorist threat assessment can be found in Chapter 6.

The special effects of catastrophic failure

The special effects of catastrophic failure or large numbers of casualties need special attention. In terrorist attacks, the number of casualties caused is often decided by whether or not a warning is given before a terrorist device is detonated; a warning allows emergency action to be taken. In the UK we have become used to one form of terrorism, Irish terrorism, which has its roots in a Christian culture and where the hysterical suicide of a terrorist bomber is not admired. These terrorists have generally wanted to exert pressure through mass publicity rather than to cause mass casualties *per se*. By contrast, terrorists from the Middle East have always sought casualties with abandon, and a group in the future may take action against our society.

Thus, if a structure is to have a useful life of several decades, the effect of an attack without prior warning must be anticipated. Consider what would have happened if the bombs in London at St Mary's Axe in 1992 (see Plates 1 and 2) or Bishopsgate in 1993 had exploded during working hours and without warning. The deaths from flying glass in the Commercial Union building could have been into the hundreds (quite apart from the many other people who would have been badly injured) and few would have escaped unhurt from the collapsed front of the Chamber of Shipping. A more important consequence in such situations could be the public perception of the effects of no warning being given and their possible reluctance to work in unprotected buildings while the potential for another attack remains. These arguments are important when trying to decide the cost/benefit of protective measures.

Partial safety factors in blast design

When designing structural elements in accordance with limit state principles, partial safety factors are applied to loads and the strengths of materials. In designing against blast loading, the following special conditions will usually apply:

(a) The incident will be an unusual event.
(b) The threat will be specified in terms of an explosive charge weight at a stand-off, which can only be an estimate and already subject to a risk assessment.

Plate 1. Devastating effects of bombing in St Mary's Axe, April 1992

    (c) For economic design, some plastic deformation is normally permitted. The level of damage is specified in terms of the limiting member deflection or support rotations.
    (d) The strengths of materials will be enhanced because of the high rate of strain to which they will be subjected.
    (e) The strengths of in situ materials often exceed the characteristic values.

For these reasons, the design initially should be carried out at the ultimate limit state, with partial safety factors for both load and materials set at 1·0. However, special enhancement factors may be applied to material strengths. These are considered in further detail in Chapter 5. A limit is usually placed on the deformation of members to permit some functionality after the event.

A design philosophy: planning for protection

A starting point for the design of a building that resists blast loading is to consider the building layout and arrangements. The aim here is to decide what needs protection (the contents or the structure itself); to imagine how damage or injury will be caused; and to consider how the building or structure can be arranged to give the best inherent protection. Specifically for protection against terrorist attacks, the building design should achieve one or all of the following:

(a) *Deflect* a terrorist attack by showing, through layout, security and defences, that the chance of success for the terrorist is small; targets that are otherwise attractive to terrorists should be made anonymous.

(b) *Disguise* the valuable parts of a potential target, so that the energy of attack is wasted on the wrong area and the attack, although

Plate 2. *Debris and destruction: Bishopsgate, April 1993*

completed, fails to make the impact the terrorist seeks; it is reduced to an acceptable annoyance.

(c) *Disperse* a potential target, so that an attack could never cover a large enough area to cause significant destruction, and thereby impact; this is suitable for a rural industrial installation, but probably unachievable for any inner-city building.

(d) *Stop* an attack reaching a potential target by erecting a physical barrier to the method of attack; this covers a range of measures from vehicle bollards and barriers to pedestrian entry controls. Against a very large car bomb, in particular, this is the only defence that will be successful.

(e) *Blunt* the attack once it reaches its target, by hardening the structure to absorb the energy of the attack and protect valuable assets.

The first three of these objectives can often be met at no cost, while the last two require extra funds or special detailing. The final objective, to blunt the attack, is the subject of the remainder of this book and involves the following procedure.

From a threat analysis, one can find the size and location of the explosion to protect against. By using the relationship that the intensity of a blast decays in relation to the cube of the distance from the explosion one can adopt an idealized blast wave at the target. Using published data (Chapter 3) the characteristics of that blast wave can be determined.

The positive phase duration of the blast wave is then compared with the natural period of response of the structure or structural element. The response is defined by the two possible extremes: *impulsive* (where the load pulse is short compared to the natural period of vibration of the structure) and *quasi-static* or *pressure* (where the load duration is long compared with the natural response time of the structure). In between is a regime where the load duration and structural response times are similar, and loading in this condition is referred to as *dynamic* or *pressure-time*.

For impulsive loading most of the deformation will occur after the blast load has finished. The impulse, which can be determined, imparts kinetic energy to the structure, which deforms and acquires strain energy. The strain energy is equivalent to the area beneath the resistance-deflection function for the structure. This function is the graph of the variation of the resistance that the structure offers to the applied loading as the displacement of the structure increases. It should be based upon the dynamic behaviour of the structure, taking due account of compressive and tensile membrane action.

For quasi-static loading, the blast will cause the structure to deform whilst the loading is still being applied. The loading does work on the structure, causing it to deform and acquire strain energy as before.

For dynamic loading, if the design cannot be covered by the two previous methods, the structural response may involve solving the equation of motion of the structural system.

Reference

1. Institution of Structural Engineers, *The structural engineer's response to bomb explosion*. (In press).

2 Basic guidelines for enhancing building resilience

Introduction
Effects of an external explosion

An explosion is a very fast chemical reaction producing transient air pressure waves called blast waves. The processes of blast wave formation and quantification are discussed in detail in Chapter 3. For a ground-level explosive device (such as a bomb in a vehicle), the pressure wave will travel away from the source in the form of a hemispherical wavefront if there are no obstructions in its path. The peak overpressure (the pressure above normal atmospheric pressure) and the duration of the overpressure vary with distance from the device. The magnitude of these parameters also depends on the explosive materials from which the bomb is made and the packaging method for the bomb. Usually the size of the bomb is given in terms of a weight of TNT. Methods exist for the design of structural elements subjected to blast loads from bombs of specified charge weights: these are discussed in Chapter 5.

City streets confine the blast wave and prevent it from radiating hemispherically and this tends to increase the pressures to which buildings are subjected. The blast pressure waves will also be reflected and refracted by buildings, travelling around the corners and curves of a building. Blast waves are very intrusive: they will travel down side streets and over the tops of buildings, and thus all sides of a building will be subject to overpressures. As the wave moves further from the source of the explosion, the peak overpressure drops. However, the confining effect of buildings, called 'funnelling', and rising ground means that the pressure drops more slowly than in open ground and buildings can be at risk at what might normally be considered safe distances (see Plate 3).

When blast waves impinge directly onto the face of a building, they are reflected from the building. The effective pressure applied to that face of the building is magnified when this occurs. If a bomb is very close to a building, the building will also be impacted by shrapnel from the bomb

Plate 3. Confining effects of narrow city streets

packaging and by debris from the break-up of 'street furniture' such as litter bins and so on (Plate 4). This shrapnel moves at high velocity and will penetrate thin building façades and unprotected glazing. This effect will be hazardous to personnel who should, if possible, have the chance to avail themselves of the protection offered by solid internal walls. Design methods are available which can mitigate this effect.

All of the above factors contribute to the variation in the effects of an explosion experienced by a particular building. For a city centre building in particular, the exact location of a bomb relative to the building is very important. Therefore, it is impossible to predict with great accuracy the effects of a bomb explosion on a particular building at the design stage. Instead, the designer should attempt to form an opinion about the possible threats and the likely effects of such threats.

Stand-off distance (the distance between the bomb and the building) is a fundamental parameter when determining the blast pressures experienced by a building. As stand-off distance increases, blast pressure drops significantly. Therefore, putting distance between the building and the bomb is extremely helpful in reducing blast effects on the building. However, that is not always a controllable parameter. For example, in a city, space is at a premium and the provision of large stand-off distances may be impossible. Indeed the question as to whether the building needs to occupy a high-risk site should be addressed. A further factor that should be noted is that the client may be unable to prevent a device being placed immediately outside the entrance to the building. However, it should be possible to take measures to maintain stand-off that would prevent, for example, a vehicle mounting the pavement outside a building, parking adjacent to the building or approaching the front entrance. The installation of bollards and other substantial items of street furniture could be considered. It may also be worth investing in 'hard' landscaping — incorporating steps or mounds adjacent to the building entrance to contribute to stand-off. However, care should be taken if considering 'soft' landscaping to ensure that concealment places for the smaller carried devices are not created by dense planting.

For a device placed inside a building (the stand-off distance being now effectively zero), greater damage and more injuries would be caused than if the same sized device were deployed outside the building. Therefore, the installation of an access control system on both pedestrian and vehicular entrances will minimize the opportunity for placement of the majority of the types of device that could be introduced into a building. This will lead to a minimization of the hazard to both people and property.

The response of a building

Loads from blasts are transient, so the ductility and natural period of vibration of the structure govern its response to a given explosion. The natural period can be calculated crudely using methods in the Uniform Building Code [1]. In general, a tall building will have a low natural frequency and thus a long response time in relation to the duration of the load. Individual elements (e.g. columns and beams) will have natural response times that may approach the loading duration. Ductile elements made of steel and reinforced concrete can absorb a lot of strain energy (i.e. they can undergo substantial bending without breaking), while

Plate 4. Street debris generated by a blast

elements made of brittle materials such as glass, brick, timber and cast-iron fail abruptly with little prior deformation.

Flexible components such as high-mass, long-span beams and floors can absorb a great deal of the energy delivered by a blast load. On the other hand, rigid, short-span lightweight elements (e.g. conventional glazing components) are poor energy absorbers and can fail catastrophically (Plate 5). This response is a function both of the material properties and the way such materials are used. For example, it is possible to use concrete to create both a flexible frame or a much more rigid bunker-like structure. Massive structures, in general, respond better than those of lightweight construction.

Plate 5. Widespread damage to conventional glazing

It is worth commenting on techniques for enhancing the resistance of glazing elements to blast loading. Such techniques do not involve the strengthening of the glazing lights alone: without proper dynamic design of the frames, the resulting assembly may perform even worse than with no treatment.

There are three main ways of providing glazing protection. These are:

(a) The application of transparent polyester anti-shatter film (ASF) to the inner surface of the glazing with the optional provision of bomb blast net curtains (BBNC).

(b) The use of blast-resistant glass.

(c) The installation of blast-resistant secondary glazing inside the (existing) exterior glazing.

Blast-resistant glazing generally consists of laminated annealed glass or laminated toughened glass or, in double-glazed units, combinations of toughened and laminated glass. In double-glazed units laminated glazing is generally the inner pane though, for preference, both should be laminated material. As noted above, a suitably designed robust frame and fixings will need to be used to install the glazing.

The primary purpose of glazing protection is to reduce the number of sharp-edged fragments that are created when ordinary annealed or toughened glass is subjected to blast loading. These shards, which travel at high speed, can cause severe injuries to personnel, can damage delicate equipment inside buildings such as the hardware for computer systems and cause major problems should they enter the air-conditioning system. Blast-resistant glazing has the capability to decrease greatly (by up to 90%) the number of loose shards produced. In addition, the amount of glass falling from the building after an explosive event is greatly reduced (Plate 6). This enables subsequent access to the building to be very much quicker and, since the majority of the glazing is attached either to the ASF or to the laminated glass polymeric 'inter-layer', clearing up is easier and quicker than without such glazing protection. A simplified procedure is presented in Appendix A for use as a guide to the design of glazing to resist blast loading.

Plate 6. Europa Hotel, Belfast: laminated glazing panels are retained in robust frames despite large deflections. Photograph courtesy of Kirk McClure Morton

As well as windows, doors and service openings are vulnerable points on a building which external blasts can penetrate. In the case of bigger explosions, the structure as a whole becomes affected. Although framed buildings generally perform better than panel or load-bearing construction, it is possible that all floors could be momentarily lifted by the blast that has entered. Blast pressures can also damage equipment within the building and travel to its heart (e.g. to plant rooms) via air ducts.

As well as incurring ear and lung injuries from the blast overpressure, occupants can suffer injury from missiles such as glass and shrapnel and from high-speed spalling of concrete cover as elements flex. Some quantification of injury levels is presented in Chapter 4. The sudden movement of the building structure can also displace features such as heavy suspended ceilings and office furnishings such as filing cabinets and desks.

The requirements of the client

The type of building being considered in this chapter is a typical commercial multi-storey office block, assumed to be a completely new structure without a retained façade. The client or architect has a number of choices to make regarding the fundamental aspects of the building design. The engineer can advise the client regarding these choices based on the points made in this chapter. However, the building will not suddenly become blast resilient by compliance with these basic rules, and the client should understand this.

The client has to decide to what expense he or she is prepared to go in order to protect the building. There are features that can be included in the design which will enhance the resilience, but these are expensive. The level of protection provided has to be decided in terms of performance under various conditions. For example, consideration should be given to the theoretical stand-off distance for a particular device at which the glazing should remain intact. Also, providing protection against a higher blast pressure from a bigger or closer device will be more expensive. There could be savings in the future if an incident were to occur, but the risk of this occurring has to be balanced against the initial expense of the resilience provisions.

The fundamental requirement is safety of the building's occupants. The Building Regulations [2] specify that a local collapse should not prejudice the overall stability of a structure. This requirement was included as a result of the collapse of the Ronan Point block in 1968 which occurred as the result of a gas explosion. This philosophy is pertinent for a terrorist attack where a large device placed close to the structure could sever one or more columns. The ability of the building to survive without collapse, except in the immediate locality of the explosion, will save the lives of occupants and assist rescue of those who are trapped. As only framed buildings can be considered to offer the ability to survive against a terrorist attack, it is therefore advised that all framed structures should be designed in accordance with the robustness clauses of BS 5950 [3] or BS 8110 [4] as a minimum.

The provision of bomb shelter areas (BSAs) is a second requirement and is considered a better option than evacuation, in many circumstances. These are areas within the building in which the occupants can seek refuge in the event of an alert and will be discussed in more detail later in this chapter.

The client will then have to decide the size of the device loading from which he or she requires the building to survive and be subsequently repairable (i.e. the building should suffer no major structural defects). This implies that a limit would need to be placed on the magnitude of plastic deformations. If the client requires survival of the building's glazing then the loading on the building would need to be kept much lower to elicit a truly resilient performance. It is worth noting, therefore, that certain design details can mean that outwardly similar buildings have dissimilar performances. Some extra expense for design enhancements can save a great deal of money in the amount of repair required.

As a consequence of this, the client must be presented with clear options regarding the required level of protection with due regard to the potential blast loading from various devices placed at various stand-off distances. The client is then in a position to assess the cost implications of these and choose the appropriate level of protection.

Architectural aspects and design features

The architect may well be wishing to design a building with notable features such as a large glazed frontage, possibly in conjunction with a glazed atrium. These are examples of features that would lead to very vulnerable structures with little inherent resistance to blast loads. They also cause high hazards from secondary fragments generated by failure of parts of the structure. If such features are to be specified, then the potential repair bill is likely to be large. Also, there could be implications for building insurance: premiums have already increased for the provision of cover against acts of terrorism and an obviously vulnerable building is likely to attract a higher premium.

Therefore, there are several aspects of the design of the façade of a building that should be considered when attempting to minimize the vulnerability of the people within the building and the damage to the building itself as described below.

(a) As a general rule, it is a good idea to minimize the amount of glazing on the façade of the building. This limits the amount of internal

damage from the glazing and the level of the blast loading that can enter the building to cause damage to fixtures and fittings.
(b) Ensure that the cladding is securely fixed to the structure with easily accessible fixings. This will allow rapid inspection and, if necessary, replacement after an event.
(c) Ensure that the cladding system allows for the easy removal and installation of individual panels. This will avoid the need to remove all the panels after an event if only one is damaged.
(d) Avoid the use of deep reveals and deep, flat window sills which are accessible from ground level as these provide ideal concealment places for small devices.
(e) Minimize the use of deep surface profiling because such features can enhance blast effects by virtue of the complex reflections produced and lead to a greater level of damage than would be produced with a plane façade.

Likewise certain structural elements already present in a building can be arranged to advantage to provide protection, e.g. a diaphragm wall otherwise designed to produce shear resistance. The building would also benefit if design takes into account the need for post-event inspection of details.

Provision of shelter areas

In order to safeguard people from the effects of a blast, the first action that must be taken is to move them as far away from the device as possible: a natural reaction might well be to evacuate personnel swiftly out to the street. However, this may not always be easily and safely accomplished. Consideration then should be given to moving people to a BSA.

The shelter concept has been employed over many centuries in times of conflict. During the Second World War shelters took many forms ranging from the arch-shaped corrugated steel Anderson shelters that were built in many back gardens and the steel Morrison tables for use within the home, to the use of London Underground stations where whole communities sought refuge from the Blitz. Thus, the provision of a BSA at the design stage of a new building or its creation within an existing building will be beneficial even though such a provision will not provide complete protection.

It is worth commenting on the BSA philosophy in the context of the present terrorist threat. The philosophy is based on the examination of the effects of blast from large vehicle bombs (in both Britain and Northern Ireland) on various structural frames. The conclusions drawn from this examination are:

(a) Structural frames of modern reinforced concrete and structural steel

Plate 7. Grand Opera House, Belfast: some load-bearing masonry walls are unstable, while others remain relatively intact — the structure was subsequently repaired

buildings remain relatively intact and do not suffer major collapse, although lightweight cladding and internal partitions and ceilings are totally destroyed.
(b) Heavy masonry fronted and internal steel-framed buildings constructed at the turn of the century also remain relatively intact (Plate 7).
(c) Internal rooms of the buildings described in (a) and (b) where walls are constructed of reinforced concrete or masonry also remain relatively intact.

On the other hand, severe personal injuries from falling glass and other debris are likely to occur at considerable distance from the seat of a large explosion. For a device comprising 1 tonne of home-made explosive (HME), injuries from falling glass may occur at up to 250 m from the explosion and, from metal fragments, at up to 500 m (Plate 8).

Plate 8. The disintegration of buildings in Bangor, N. Ireland due to blast loading produces fragments that could inflict serious injuries. Photograph courtesy of Kirk McClure Morton

The conclusion to be drawn from this evidence is that moving staff to areas within a building rather than evacuating onto the streets in the event of a bomb alert is the best policy for very large vehicle bombs as well as other kinds of smaller externally deployed devices. For buildings of the types described above, BSAs should be located:

- remote from windows, external doors and walls
- remote from the 'perimeter structural bay' (i.e. that part of the floor structure, at all levels, between the building's perimeter and the first line of supporting columns)
- in areas surrounded by full-height masonry or concrete walls (e.g. internal corridors, toilet areas or conference rooms).

BSAs should not be located in stairwells or in areas with access to liftshafts as these generally open out at ground floor level directly to the street. However, in certain buildings where the stair and lift cores are totally enclosed, a very good BSA may be established.

It is important to have adequate communications established within designated BSAs to inform staff of any subsequent action which may need to be taken — e.g. to remain within the BSA, to move to another BSA (if the location of a device presents a particular threat), to evacuate the building or to give the 'all clear'. The system of communication may be a public address system (which may require stand-by power should an explosion cut the main electricity supply), hand-held radio transmitter/receivers or other stand-alone audio-communication links. Further information on such contingency planning may be found in Chapter 6.

Location of key equipment and services

The most essential assets for the client (e.g. central computing facilities) should be safeguarded and the option for back-up facilities should be considered. Operational resilience could involve having duplicate critical systems either located in another building situated at least 1 kilometre away or, if it is large enough, located at the opposite end of the building under threat.

If, however, a critical asset cannot be duplicated then blast effects should be minimized by locating it in internal and relatively safe rooms that should, ideally, offer the same level of protection provided by a BSA in the building.

Building form

The essential requirement is for a ductile structure to resist the worst effects of blast loading. This requirement forces the designer to consider construction of reinforced concrete or steel-framed buildings (Plate 9). A framed building, tied together adequately, will have many different load

Plate 9. Comparative blast damage to steel framed flats and a church with load-bearing walls. Photograph courtesy of Francis Walley

paths by which to transfer loads to the ground. If this aspect is coupled with ductile behaviour, the structure will fulfil its primary safety requirement. As noted above, internal partitions can be a serious hazard to safety. They are generally lightweight and can be easily demolished under relatively low levels of blast loading and would cause a secondary flying debris hazard. Careful detailing to either remove or provide sufficient strength for partitions will remove the hazard. Floors should be tied to the frame, and have the potential to withstand stress reversals. Blast loads can often impart uplift pressures that are sufficient to overcome gravity loads on a floor. Unless the floor is tied down, it will dislodge. Floors can act as diaphragms which transfer lateral loads between frames, and the loss of these floors could initiate a progressive collapse. Suitable measures to counter these effects are the provision of continuous spans and, particularly in lower floors, the introduction of reinforcement in both faces of the slab (e.g. mesh mats should be cast in the top of the slab).

Design of individual structural elements

Just as with structures designed for static loading, the engineer should use limit state design techniques: ultimate limit state (ULS) (collapse) and serviceability limit state (SLS) (functionality after event) approaches.

The collapse state requires sufficient ductility to dissipate the blast energy without causing collapse. Beams should have symmetrical primary reinforcement and sufficient shear links to provide restraint and prevent premature shear failure. Deflections are usually allowed to exceed the elastic limit by up to a certain factor, known as the ductility ratio. However, it is not envisaged that a dynamic design calculation need always be undertaken by structural engineers designing an office building. In some cases, by designing in accordance with BS 5950 [3] or BS 8110 [4, 5], a satisfactory design may be achieved as long as moment-resisting structures with continuous spans are detailed. It should be noted, however, that these standards have not been prepared to deal specifically with terrorist explosions. For example, the minimum horizontal load to be applied to the exterior of a building is not applicable and the pressure of 34 kN/m^2 specified for the design of key elements has been based on internal domestic gas explosions.

In other cases it will be necessary to design elements specifically to resist the blast loading. The design of structural elements to resist blast loading at the ULS is covered in some depth in Chapter 5. Bearing in mind that the cost of the structural frame in a building may represent only 25–30% of total building costs, there is evidence [6] to show that the required increases in element size may result in an increase in cost of no more than 2–3% overall.

The performance of the connections is vital to the behaviour at ULS (Plate 10). In steel-framed construction, particularly, the joints must be

Plate 10. Failure of beam-column joint in reinforced concrete frame due to poor detailing

detailed to carry moments around the frame. The joints should be capable of resisting stress reversals, as should the structure as a whole. A column-beam joint may not transfer moment into the column in static design because the moment at each beam end equalizes. However, in a blast situation which could impose large sway deflections on the structure and generate substantial moments, these moments have to be transferred to the columns from the beams.

At ULS, the frame will survive, but may be so badly deformed as to be suitable only for demolition. Therefore, limitations are often placed upon the magnitude of material stresses and member deflections to provide some level of post-event functionality. A further SLS requirement is the maintenance of windows, cladding and internal fixings. These should have a measure of blast resilience; flexible or ductile partitions tied to the structural frame are therefore required. Windows are to be laminated, or have ASF applied (as described above), if a reasonable level of resistance is to be provided in a moderate risk area.

Plate 11. Concrete framed structure with concrete cladding panels, whose fixings have failed either partially or completely

Serviceability design could entail the design of details that enable straightforward repair of the damaged fenestration. Cladding may be allowed to yield, so long as fixings do not fail (Plate 11). The building should remain weathertight, and repair could be effected without major interruption of the operation of the office.

Building layout

The layout of the building can enhance the performance of the structure when subjected to blast loading and enable the provision of BSAs. Such performance enhancements can be achieved by adherence to some or all of the following guidelines:

(a) A building should be three bays wide at least, to allow personnel

to move away from windows and into the relative safety of a central corridor or core area.
(b) Buildings should have structural core areas, preferably formed in reinforced concrete, for use as BSAs.
(c) Re-entrant corners should be avoided since they increase blast pressures locally because of the complex reflections created. (See also comments on building façades above.)
(d) Reductions in blast loads can be achieved by, for example, setting back upper floors to increase the stand off. This can only be effective for relatively low buildings with a large floor area.

References

1. ICBO. *Uniform Building Code*. International Conference for Building Officials, Whittier, USA, 1985.
2. HMSO. *The Building Regulations*. HMSO, London, 1985.
3. British Standards Institution. *Structural use of steelwork in building. Code of practice for design in simple and continuous construction: hot rolled sections*. BSI, London, 1985, BS 5950: Part 1.
4. British Standards Institution. *Structural use of concrete. Code of practice for design and construction*. BSI, London, 1985 (amended 1993), BS 8110: Part 1.
5. British Standards Institution. *Structural use of concrete. Code of practice for special circumstances*. BSI, London, 1985, BS 8110: Part 2.
6. Elliott C.L., Mays G.C. and Smith P.D. The protection of buildings against terrorism and disorder. *Proc. Instn Civ. Engrs, Structures and Buildings*, 1992, **94**, 287–297. Discussion, 1994, **104**, 343–350.

3 Blast loading

Notation

a_0	speed of sound in air at ambient conditions
A	area of target loaded by blast
A_s	total inside surface area of structure
b	wavefront parameter
B	target dimensions
C_r	reflection coefficient
C_D	drag coefficient
d	charge diameter
F_D	drag force
H	target dimensions
i_g	gas pressure impulse
i_s	specific side-on impulse
i_r	specific reflected impulse
i^-	negative phase specific impulse
p	pressure
p_0	atmospheric pressure
p_{QS}	peak quasi-static gas pressure
p_r	peak reflected overpressure
p_s	peak side-on overpressure
p_{stag}	stagnation pressure
Δp_{min}	peak underpressure
q_s	peak dynamic pressure
Q	mass specific energy of condensed high explosive
R	range from charge centre
S	target dimensions
t	time
t'	pressure reduction time
t_a	arrival time of blast wave front
t_{max}	blowdown time for internal explosion
t_r	reverberation time
T_s	positive phase duration
T_r	positive phase duration of reflected wave
u_s	particle velocity behind blast wave front

U_s	blast wave front speed
$\bar{u}$	Mach number of wavefront particle velocity
$\bar{U}$	Mach number of wavefront
V	volume
W	mass of spherical TNT charge
Z	scaled distance
α_e	ratio of vent area to wall area
α_I	angle of incidence
γ	specific heat ratio
λ	scale factor

Introduction

This chapter deals with the formation and quantification of blast waves produced by condensed high explosives. In particular it presents information allowing the reader to evaluate peak overpressures and the associated impulses for a range of explosives expressed in terms of a scaled distance based on range and the mass of TNT equivalent to the actual explosive being considered for both spherical and hemispherical charges. The pressure produced by the processes of reflection are evaluated and the forces that result on structures are described both for explosives external to a building and for explosions inside a structure. In the latter case, the effect of the so-called gas pressure loading produced by the products of detonation is quantified.

Explosions

Explosions can be categorized as physical, nuclear or chemical events. Examples of physical explosions include the catastrophic failure of a cylinder of compressed gas, the eruption of a volcano or the violent mixing of two liquids at different temperatures. In a nuclear explosion the energy released arises from the formation of different atomic nuclei by the redistribution of the protons and neutrons within the interacting nuclei.

A chemical explosion involves the rapid oxidation of fuel elements (carbon and hydrogen atoms) forming part of the explosive compound. The oxygen needed for this reaction is also contained within the compound so that air is not necessary for the reaction to occur. To be useful, a chemical explosive must only explode when required to and should be inert and stable. The rate of reaction (much greater than the burning of a fuel in atmospheric air) will determine the usefulness of the explosive material for practical applications. Most explosives in common use are 'condensed': they are either solids or liquids. When the explosive is caused to react it will decompose violently with the evolution of heat and the production of gas. The rapid expansion of this gas results in the generation of shock pressures in any solid material with which the explosive is in contact or blast waves if the expansion occurs in a medium such as air.

Although far less commonly deployed by terrorists, mention should be made of 'fuel-air' or 'vapour cloud' explosions which can produce damage to structures commensurate with that produced by condensed high explosives. In such events the chemical reaction is generally deflagrative (see below) producing pressures that may not be as high at a given range from the explosion centre as for an equivalent quantity of condensed explosive but which remain at a significant level for a longer period.

Explosion processes terminology

Combustion is the term used to describe any oxidation reactions, including those requiring the presence of oxygen from outside as well as those that use oxygen which is an integral part of the reacting compound.

In the case of explosive materials which decompose at a rate much below the speed of sound in the material, the combustion process is known as deflagration. Deflagration is propagated by the liberated heat of reaction: the flow direction of the reaction products is in opposition to the direction of decomposition.

Detonation is the explosive reaction which produces a high intensity shock wave. Most explosives can be detonated if given sufficient stimulus. The reaction is accompanied by large pressure and temperature gradients at the shock wave front and the reaction is initiated instantaneously. The reaction rate, described by the detonation velocity, lies between about 1500 and 9000 m/s which is appreciably faster than propagation by the thermal processes active in deflagration. It is worth noting that, if a condensed high explosive is detonated in contact with a structure, the impact of the detonation wave produces a shattering effect on the material of the structure known as 'brisance'.

Explosives classification

High explosives detonate to create shock waves, burst or shatter materials in or on which they are located, penetrate materials, produce lift and heave of materials and, when detonated in air or under water, produce air-blast or underwater pressure pulses. Low explosives deflagrate to produce pressure pulses generally of smaller amplitude and longer duration than high explosives. Examples include propellants for launching projectiles and explosive mixtures such as gunpowder.

Classification of these materials is generally on the basis of their sensitivity to initiation. A primary explosive is one that can be easily detonated by simple ignition from a spark, flame or impact. Materials such as mercury fulminate, and lead azide are primary explosives. They are the type of materials that can be found in the percussion cap of firearm ammunition. Secondary explosives can be detonated, although less easily than primary explosives. Examples include TNT and RDX (also known as cyclonite) among many others. In firearm ammunition, a secondary explosive would be used for the main explosive charge of the shell or cartridge.

In order to achieve the required properties of safety, reliability and performance (also paying due regard to economic considerations), it is common practice in both military and commercial explosives manufacture to blend explosive compounds. For commercial use, explosives are generally made from cheaper ingredients: TNT or nitroglycerine might be mixed with low-cost nitrates, for instance. Such material has a generally short shelf-life. Military explosives are composed of more expensive ingredients (such as binary mixtures of stable compounds like TNT and RDX or HMX with TNT) and generally have a long shelf-life.

Terrorist organizations typically only have limited quantities of military-style high explosives such as Semtex and as a consequence often manufacture their own explosive materials, from farm fertilizer, for instance. In the quantities necessary to attack substantial structures, this material behaves as a detonating high explosive material.

Blast waves in air from condensed high explosives

When a condensed high explosive is initiated the following sequence of events occurs. Firstly, the explosion reaction generates hot gas which can be at a pressure from 100 up to 300 kilobar and at a temperature of about 3000–4000°C. A violent expansion of this gas then occurs and the surrounding air is forced out of the volume it occupies. As a consequence a layer of compressed air — the blast wave — forms in front of this gas containing most of the energy released by the explosion. As the gas expands its pressure falls to atmospheric pressure as the blast wave moves outwards from the source. The pressure of the compressed air at the blast wavefront also falls with increasing distance. Eventually, as the gas continues to expand it cools and its pressure falls a little below atmospheric pressure. This 'overexpansion' is associated with the momentum of the gas molecules. The result of overexpansion is a reversal of flow towards the source driven by the small pressure differential between atmospheric conditions and the pressure of the gas. The effect on the blast wave shape is to induce a region of 'underpressure' (i.e. pressure is below atmospheric pressure) which is the 'negative phase' of the blast wave. Eventually the situation returns to equilibrium as the motions of the air and gas pushed away from the source cease.

Blast wavefront parameters

Of particular importance are the blast wavefront parameters. Analytical solutions for these quantities were first given by Rankine and Hugoniot in 1870 [1] to describe normal shocks in ideal gases and are available in a number of references such as Liepmann and Roshko [2]. The equations for blast wavefront velocity, U_s, and the maximum dynamic pressure, q_s, are given below

$$U_s = \sqrt{\frac{6p_s + 7p_0}{7p_0}} \cdot a_0 \qquad (3.1)$$

$$q_s = \frac{5p_s^2}{2(p_s + 7p_0)} \qquad (3.2)$$

where p_s is peak static overpressure at the wavefront, p_0 is ambient air pressure and a_0 is the speed of sound in air at ambient pressure.

The analysis due to Brode [3] leads to the following results for peak static overpressure in the near field (when p_s is greater that 10 bar) and in the medium to far field (p_s between 0·1 and 10 bar)

$$p_s = \frac{6\cdot 7}{Z^3} + 1 \text{ bar} \qquad (p_s > 10 \text{ bar}) \qquad (3.3)$$

$$p_s = \frac{0\cdot 975}{Z} + \frac{1\cdot 455}{Z^2} + \frac{5\cdot 85}{Z^3} - 0\cdot 019 \text{ bar} \quad (0\cdot 1 < p_s < 10 \text{ bar}).$$

Here Z is scaled distance given by

$$Z = R/W^{1/3} \qquad (3.4)$$

where R is the distance from the centre of a spherical charge in metres and W is the charge mass expressed in kilograms of TNT.

The use of TNT as the 'reference' explosive in forming Z is universal. The first stage in quantifying blast waves from sources other than TNT is to convert the actual mass of the charge into a TNT equivalent mass. The simplest way of achieving this is to multiply the mass of explosive by a conversion factor based on its specific energy and that of TNT. Conversion factors for a number of explosives are shown in Table 3.1 adapted from Baker et al. [4]. From the table it can be seen that a 100 kg charge of RDX converts to 118·5 kg of TNT since the ratio of the specific energies is 5360/4520 (=1·185).

An alternative approach described in [5] makes use of two conversion factors. The choice of which to use depends on whether the peak overpressure or the impulse delivered is to be matched for the actual explosive and the TNT equivalent. Thus, for Compound B the equivalent pressure factor is 1·11 while that for impulse is 0·98.

The TNT equivalence of terrorist-manufactured explosive material (known as home-made explosive or HME) is difficult to define precisely because of the variability of its formulation and the quality of the control used in its manufacture. TNT-equivalent factors ranging from as low as 0·4 (for poor quality HME) up to almost unity have been suggested.

Table 3.1. Conversion factors for explosives

Explosive	Mass specific energy Q_x (kJ/kg)	TNT equivalent (Q_x/Q_{TNT})
Compound B (60% RDX 40% TNT)	5190	1·148
RDX (Cyclonite)	5360	1·185
HMX	5680	1·256
Nitroglycerin (liquid)	6700	1·481
TNT	4520	1·000
Blasting Gelatin (91% nitroglycerin, 7·9% nitrocellulose, 0·9% antacid, 0·2% water)	4520	1·000
60% Nitroglycerin dynamite	2710	0·600
Semtex	5660	1·250

Similarly for fuel-air or vapour cloud explosions, TNT equivalence is hard to specify accurately though a factor of between 0·4 and 0·6 is sometimes used.

Other important blast wave parameters

Other significant blast wave parameters include T_s, the duration of the positive phase (the time when the pressure is in excess of ambient pressure) and i_s the specific impulse of the wave which is the area beneath the pressure-time curve from arrival at time t_a to the end of the positive phase as given by

$$i_s = \int_{t_a}^{t_a+T_s} p_s(t)\, dt \tag{3.5}$$

A typical pressure-time profile for a blast wave in free air is shown in Fig. 3.1, where Δp_{min} is the greatest value of underpressure (pressure below ambient) in the negative phase of the blast. This is the rarefaction or underpressure component of the blast wave. Brode's solution for Δp_{min} (bar) is

$$\Delta p_{min} = -\frac{0\cdot 35}{Z} \quad (Z > 1\cdot 6) \tag{3.6}$$

and the associated specific impulse in this phase i^- is given by

$$i^- \approx i_s\left[1 - \frac{1}{2Z}\right] \tag{3.7}$$

A convenient way of representing significant blast wave parameters is to plot them against scaled distance as shown on Fig. 3.2 which is adapted

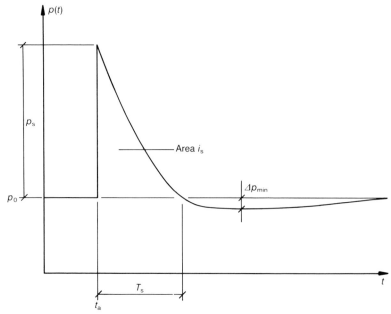

Fig. 3.1. Typical pressure–time profile for blast wave in free air (10)

from graphs presented in a number of references such as Baker *et al.* [4] and the design code TM5-1300 [6].

The directory of significant blast-wave parameters should also include dynamic pressure, q_s, blast-wave front speed U_s expressed as $\bar{U}\,(=U_s/a_0)$, particle velocity just behind the wavefront, u_s, expressed as $\bar{u}\,(=u_s/a_0)$ and the waveform parameter b in the equation describing the pressure-time history of the blast wave [(eqn (3.10)]. Figure 3.3 shows these parameters plotted against scaled distance Z.

Blast wave scaling laws

The most widely used approach to blast-wave scaling is that formulated independently by Hopkinson [7] and Cranz [8]. Hopkinson-Cranz scaling is commonly described as cube-root scaling. Thus, if the two charge masses are W_1 and W_2 of diameter d_1 and d_2, respectively then, for the same explosive material, since W_1 is proportional to d_1^3 and W_2 is proportional to d_2^3, it follows that

$$\frac{d_1}{d_2} = \left(\frac{W_1}{W_2}\right)^{1/3} \qquad (3.8)$$

Therefore, if the two charge diameters are in the ratio $d_1/d_2 = \lambda$, then if the same overpressure p_s is to be produced from the two charges the

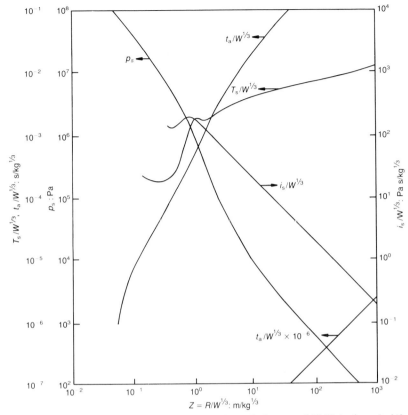

Fig. 3.2. *Side-on blast wave parameters for spherical charges of TNT in free air [4]*

ratio of the ranges at which the particular overpressure is developed will also be λ, as will the positive phase duration ratio and the impulse ratio. Ranges at which a given overpressure is produced can thus be calculated using the results of eqn (3.8). For example

$$\frac{R_1}{R_2} = \left(\frac{W_1}{W_2}\right)^{1/3} \qquad (3.9)$$

where R_1 is the range at which a given overpressure is produced by charge W_1 and R_2 is the range at which the same overpressure is generated by charge W_2.

The Hopkinson-Cranz approach leads readily to the specification of the scaled distance Z ($=R/W^{1/3}$) introduced above: it is clear that Z is the constant of proportionality in relationships such as those of eqn (3.8). The use of Z in Figs 3.2, 3.3 and 3.5 allows a compact and efficient presentation of blast wave data for a wide range of situations.

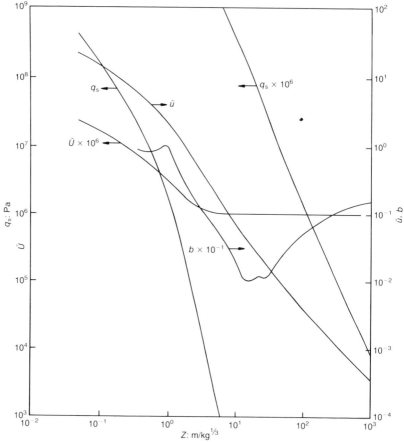

Fig. 3.3. Additional blast wave parameters for spherical charges of TNT in free air [4]

Hemispherical surface bursts

The foregoing sections refer to free-air bursts remote from any reflecting surface and are usually categorized as spherical airbursts. When attempting to quantify overpressures generated by the detonation of high explosive sources in contact with the ground, modifications must be made to charge weight before using the graphs presented earlier.

Good correlation for hemispherical surface bursts of condensed high explosives with free air burst data results if an enhancement factor of 1·8 is assumed. In other words, surface bursts produce blast waves that appear to come from free air bursts of 1·8 times the actual source energy. It should be noted that, if the ground were a perfect reflector and no energy was dissipated (in producing a crater and groundshock) the reflection factor would be 2 (Plate 12).

Plate 12. Bishopsgate crater generated by vehicle bomb

Blast wave pressure profiles

The pressure-time history of a blast wave is often described by exponential functions such as the Friedlander equation [in which b is called the waveform parameter (see above)].

$$p(t) = p_s\left[1 - \frac{t}{T_s}\right] \exp\left\{-\frac{bt}{T_s}\right\} \qquad (3.10)$$

For many purposes, however, approximations are quite satisfactory. Thus, linear decay is often used in design where a conservative approach would be to represent the pressure-time history by line I in Fig. 3.4. Alternatively it might be desirable to preserve the same impulse in the idealized wave shape compared with the real profile as illustrated by line II in Fig. 3.4 where the areas beneath the actual decay and the approximation are equal.

Blast wave interactions

When blast waves encounter a solid surface or an object made of a medium more dense than air, they will reflect from it and, depending on its geometry and size, diffract around it. The simplest case is that of

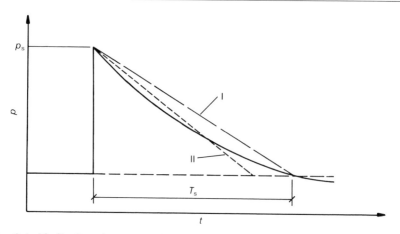

Fig. 3.4. Idealization of pressure–time profile [10]

an infinitely large rigid wall on which the blast wave impinges at zero angle of incidence. In this case the incident blast wave front, travelling at velocity U_s, undergoes reflection when the forward moving air molecules in the blast wave are brought to rest and further compressed inducing a reflected overpressure on the wall which is of higher magnitude than the incident overpressure.

Rankine and Hugoniot derived the equation for reflected overpressure p_r (assuming that air behaves as a real gas with specific heat ratio $C_p/C_v = \gamma$) in terms of incident peak overpressure and dynamic pressure [given by eqn (3.2)] as

$$p_r = 2p_s + (\gamma + 1)q_s \tag{3.11}$$

Substitution of q_s into this equation gives

$$p_r = 2p_s \left[\frac{7p_0 + 4p_s}{7p_0 + p_s} \right] \tag{3.12}$$

when, for air, γ is set equal to 1·4. If a reflection coefficient C_r is defined as the ratio of p_r to p_s then inspection of this equation indicates an upper and lower limit for C_r. When the incident overpressure p_s is a lot less than ambient pressure (e.g. at long range from a small charge) the lower limit of C_r is 2. When p_s is much greater than ambient pressure (e.g. at short range from a large charge) we have an upper limit for C_r of 8. However, because of gas dissociation effects at very close range, measurements of C_r of up to 20 have been made. Figure 3.5 shows reflected overpressure and impulse i_r for normally reflected blast wave parameters plotted against scaled distance Z.

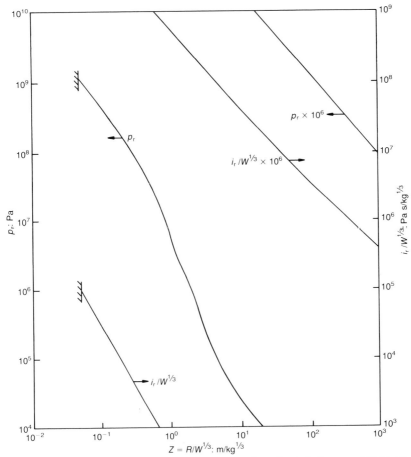

Fig. 3.5. Normally reflected blast wave parameters for spherical charges of TNT [4]

Regular and Mach reflection

In the discussion above, the angle of incidence α_I of the blast wave on the surface of the target structure was zero. When α_I is 90° there is no reflection and the target surface is loaded by the peak static overpressure which is sometimes referred to as 'side-on' pressure. Regular reflection occurs for angles of incidence from 0° up to approximately 40° in air after which Mach reflection takes place. Figure 3.6 shows the concepts of side-on pressure, regular reflection at both 0° angle of incidence (often called 'face-on' loading) and at α_I between 0° and 40° together with Mach reflection (when α_I exceeds 40°). Figure 3.7 shows reflection coefficient C_r plotted against α_I for a range of incident overpressures. Note that the Rankine–Hugoniot prediction of a maximum reflection coefficient of 8 is clearly exceeded at higher values of p_s.

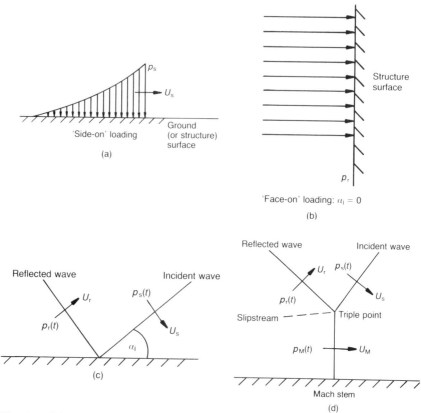

Fig. 3.6. *Side-on and face-on pressure loading, regular and Mach reflection*

As noted above, the Mach reflection process occurs when α_I exceeds about 40° in air. Mach reflection is a complex process and is sometimes described as a 'spurt'-type effect where the incident wave 'skims' off the reflecting surface rather than 'bouncing' as is the case at lower values of α_I. The result of this process is that the reflected wave catches up with and fuses with the incident wave at some point above the reflecting surface to produce a third wave front called the Mach stem. The point of coalescence of the three waves is called the triple point. In the region behind the Mach stem and reflected waves is a slipstream region where, although pressure is the same, different densities and particle velocities exist. The formation of a Mach stem is important when a conventional device detonates at some height above the ground and also occurs when a device is detonated inside a structure where the angles of incidence of the blast waves on the internal surfaces can vary over a wide range.

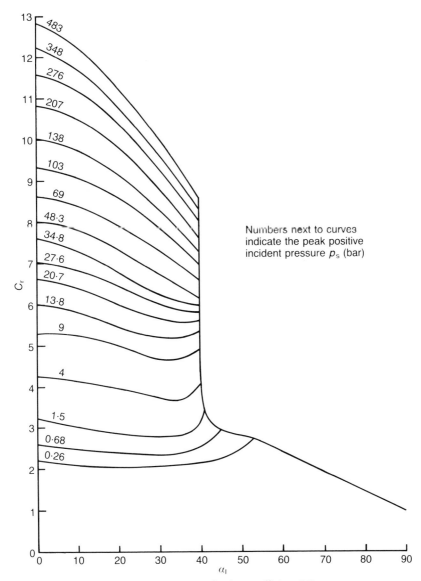

Fig. 3.7. *Effect of angle of incidence on reflection coefficient [6]*

Blast wave external loading on structures

The foregoing discussion is centred on reflecting surfaces that are essentially infinite and do not allow diffraction to occur. In the case of finite target structures three classes of blast wave-structure interaction can be identified.

The first of these is associated with a large-scale blast wave: here the target structure is engulfed and crushed by the blast wave. There will also be a translational force tending to move the whole structure laterally (a drag force) but because of the size and nature of the structure it is unlikely actually to be moved: this is diffraction loading.

The second category is where a large scale blast wave interacts with a small structure such as a vehicle. Here the target will again be engulfed and crushed. There will be a more or less equal 'squashing' overpressure acting on all parts of the target and any resultant translatory force will only last for a short time. However, more significantly, a translational force due to dynamic or drag loading will act for sufficiently long to move the target and it is likely that a substantial part of the resulting damage will be as a consequence of this motion.

Finally, consider the case of a blast wave produced by the detonation of a relatively small charge loading a substantial structure. Here, the response of individual elements of the structure needs to be analysed separately since the components are likely to be loaded sequentially.

For the first and second situations above, consider the load profile for each structure with reference to Fig. 3.8. Each experiences two simultaneous components of load. The diffraction of the blast around the structure will engulf the target and cause a normal squashing force on every exposed surface. The structure experiences a push to the right as the left-hand side of the structure is loaded followed closely by a slightly lower intensity push to the right as diffraction is completed. The drag loading component causes a push on the left side of the structure followed by a suction force on the right-hand side as the blast wave dynamic pressure (the blast wind) passes over and around the structure.

With reference to Fig. 3.8 which shows the 'squashing' and dynamic pressure variation at significant times on the structure, the following points should be noted.

In Fig. 3.8(a) the peak pressure experienced by the front face of the target at time t_2 will be the peak reflected overpressure p_r. This pressure will then decay in the time interval $(t' - t_2)$ because the pressure of the blast wave passing over the top of the structure and round the sides is less than p_r (the peak top and side overpressure will be p_s). Thus, decay in front face overpressure continues until the pressure is equal to the stagnation pressure $p_{stag}(t)$ which is the sum of the time-varying static and dynamic pressures. The time t' is given approximately by

$$t = 3 \times S/U_s \tag{3.13}$$

where S is the smaller of $B/2$ or H where B is structure breadth, H is height and U_s is the blast front velocity. In Fig. 3.8(b) the deviation from the linear decay of pressure on top and sides after time t_3 is due to the complex vortices formed at the intersection of the top and the sides with

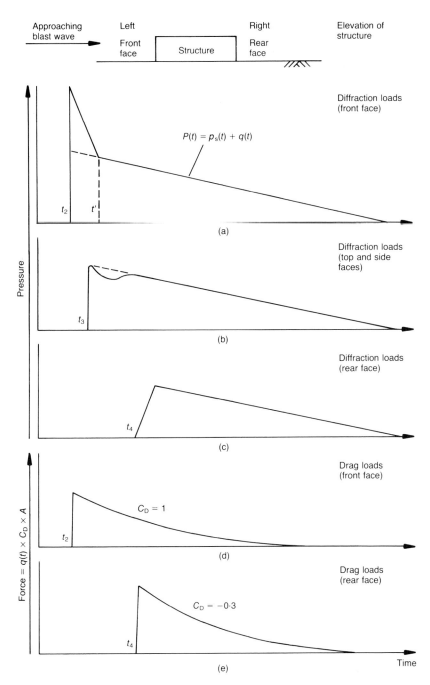

Fig. 3.8. Blast wave external loading on structures [10]

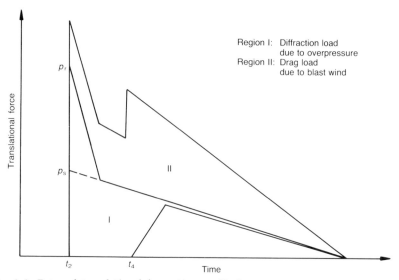

Fig. 3.9. External translational force—time profile for a structure [10]

the front. In Fig. 3.8(c) the load profile on the rear face is of finite rise time because of the time required by the blast wave to travel down the rear of the target to complete the diffraction process. Figures 3.8(d) and (e) show the forces exerted on the front and rear faces of the target by the 'blast wind' forces. The resulting drag force F_D is given by

$$F_D = C_D \times q_s(t) \times A \quad (3.14)$$

where A is the area loaded by the pressure and C_D is the drag coefficient of the target which depends on target geometry.

Combining the loading from both diffraction and drag components gives the overall translatory force-time profile as shown in Fig. 3.9.

If the target is relatively small (having only short sides) the interval (t_4-t_2) is small and area I in Fig. 3.9 is small, while area II is proportionately bigger. This loading is characteristic of a drag target.

Internal blast loading of structures

When an explosion occurs within a structure it is possible to describe the structure as either unvented or vented. An unvented structure would need to be stronger to resist a particular explosion than a vented structure where some form of pressure relief would be activated (e.g. by breaking of windows, etc.).

The detonation of a condensed high explosive inside a structure produces two loading phases. Firstly, reflected blast overpressure is generated and, because of the confinement provided by the structure, re-reflection will occur. This process will produce a train of blast waves of decaying amplitude. While this is happening the second loading phase develops as the gaseous products of detonation independently cause a build-up of pressure: this is called gas pressure loading. The load profile for the structure is likely to be complex.

The provision of venting in buildings may be beneficial for the protection of the structure against the build-up of potentially damaging gas pressures. However, for the protection of personnel, venting offers little advantage because injuries will be associated with the initial blast wave.

It is fairly straightforward to estimate the magnitude of the initial reflected blast wave parameters (p_r, i_r) by using the scaled distance curves shown in Fig. 3.5. Quantification of the re-reflected waves is generally more difficult particularly in the situations where Mach stem waves are produced. However, it is possible to undertake an approximate analysis of internal pressure-time histories by making some simplifying assumptions by approximating the pressure pulses of both incident and reflected waves as being triangular in shape with pressure-time history given by

$$p_r(t) = p_r\left(1 - \frac{t}{T_r}\right) \quad (3.15)$$

where T_r is the equivalent positive phase duration of the reflected wave. The area under the pressure–time curve for the actual pulse is the specific impulse i_r and this is set equal to the area beneath the equivalent triangular pulse. Thus, if the actual reflected specific impulse is i_r then

$$i_r = \tfrac{1}{2}T_r p_r \quad (3.16)$$

and

$$T_r = 2i_r/p_r \quad (3.17)$$

To quantify subsequent reflections the approach suggested by Baker et al. [4] is to assume that the peak pressure is halved on each re-reflection. Hence the impulse is also halved if duration of each pulse is considered to remain constant. After three reflections, the pressure of any reflected wave is assumed to be zero. With reference to Fig. 3.10, the situation can be described thus

$$p_{r_2} = \tfrac{1}{2}p_{r_1}, \quad p_{r_3} = \tfrac{1}{2}p_{r_2} = \tfrac{1}{4}p_{r_1}, \quad p_{r_4} = 0$$
$$i_{r_2} = \tfrac{1}{2}i_{r_1}, \quad i_{r_3} = \tfrac{1}{2}i_{r_2} = \tfrac{1}{4}i_{r_1}, \quad i_{r_4} = 0 \quad (3.18)$$

where it is assumed that positive phase durations remain unchanged for each reflection.

In Fig. 3.10 the reverberation time — the time delay between each blast wave arriving at the structure internal surface — is assumed constant at t_r (= $2t_a$ where t_a is arrival time of the first blast wave at the reflecting

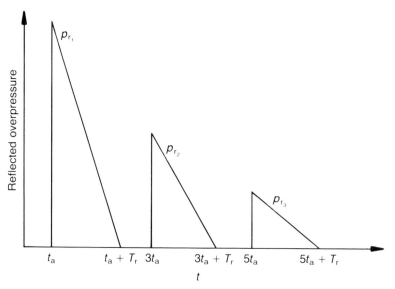

Fig. 3.10. Simplified internal blast wave reflections [4]

surface). This assumption is not strictly true because successive shocks will be weaker and so will travel slower than the first.

A further simplification suggested in [4] can be made particularly if the response time of the structure is much longer than the total load duration $(5t_a + T_r)$ (see Chapter 4) when all three pulses may be combined into a single pulse having 'total' peak pressure p_{rT} delivering a total specific impulse i_{rT}. Thus

$$p_{rT} = p_{r_1} + p_{r_2} + p_{r_3} = 1\cdot75\, p_{r_1}$$
$$i_{rT} = i_{r_1} + i_{r_2} + i_{r_3} = 1\cdot75\, i_{r_1} \qquad (3.19)$$

These approximations can be justified in that, when assessing the response of a structure, the use of the approximate input will lead to an overestimate of response leading to a conservative design.

While the reverberating blast waves are decaying the gas pressure load is developing. Its magnitude at a particular time will depend on the volume of the structure, the area of any vents in the structure and the characteristics of the particular explosive. A typical pressure-time history for a structure with some form of venting is shown in Fig. 3.11.

The figure shows a series of reverberating blast waves [approximately three in number, confirming the validity of the approach of eqn (3.18)] and a developing gas pressure load which peaks at point B and then decays. Reference [4] presents an approach allowing quantification of the important features of the history by use of a simplified form of the gas pressure component of the record. An approximate equation describing the pressure-time history of the gas pressure decay is

$$p(t) = (p_{QS} + p_0) e^{(-2.13\tau)} \qquad (3.20)$$

where p_{QS} is peak quasi-static pressure, p_0 is ambient pressure and

$$\tau = \frac{\alpha_e A_s t a_0}{V} \qquad (3.21)$$

where α_e is the ratio of vent area to wall area, A_s is the total inside surface area of the structure, V is the structure volume and a_0 is the speed of sound at ambient conditions. This equation is valid for the part of the history showing decaying pressure. The rise of gas pressure is assumed to be linear and peaks at a time corresponding to the end of the reverberation phase $(5t_a + T_r)$. The gas pressure history is shown by the dashed line in the figure. The area under the curve (ignoring the initial linear rise) is termed the gas impulse i_g which can be written

$$i_g = \int_0^{t_{max}} (p(t) - p_0) \, dt = \frac{p_1}{C} [1 - e^{-Ct_{max}}] - p_0 t_{max} \qquad (3.22)$$

in which $p_1 = p_{QS} + p_0$ and

$$C = 2.13 \alpha_e A_s a_0 / V \qquad (3.23)$$

From experimental data from several sources (for example [9]) the curves of Fig. 3.12 adapted from [4] and [5] have been shown to give reasonable predictions of peak quasi-static pressure, 'blowdown' time (t_{max}) and gas pressure impulse.

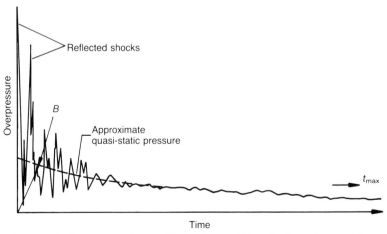

Fig. 3.11. Typical pressure–time profile for internal blast loading of a partially-vented structure [5]

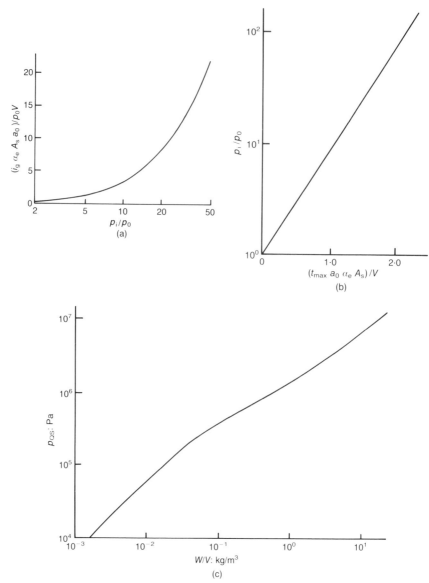

Fig. 3.12. Prediction of gas pressure impulse (i_g), 'blowdown' time (t_{max}) and peak quasi-static pressure (p_{QS}) [4]

Conclusions

This chapter has provided a summary of the methods of blast load quantification from both external and internal explosions (more details about which can be obtained from reference [10]) and allows the designer to obtain the pressure and impulse values necessary to allow the design process described in Chapter 5 to be used to produce a structure capable of resisting these loads. It should be noted that the level of damage suffered by a structure cannot be determined solely from knowledge of the pressure and impulse values from a particular explosion. It is also important to know the characteristics of the blast-loaded building, in particular the dynamic properties of the materials of construction and the form of the structure. Interdependence of loading and building characteristics is discussed in Chapter 4.

References

1. Rankine W.J.H. *Phil. Trans. Roy. Soc.*, 1870, **160**, 277–288.
2. Liepmann H.W. and Roshko A. *Elements of Gas Dynamics*. John Wiley, New York, 1957.
3. Brode H.L. Numerical solution of spherical blast waves. *J. App. Phys.*, 1955, No. 6, June.
4. Baker W.E., Cox P.A., Westine P.S., Kulesz J.J. and Strehlow R.A. *Explosion Hazards and Evaluation*. Elsevier, 1983.
5. U.S. Department of the Army Technical Manual, TM5-855-1. *Fundamentals of protective design for conventional weapons*. Washington DC, 1987.
6. U.S. Department of the Army Technical Manual, TM5-1300. *Design of structures to resist the effects of accidental explosions*. Washington DC, 1990.
7. Hopkinson B. British ordnance board minutes 13565, 1915.
8. Cranz C. *Lehrbuch der Ballistik*. Springer, Berlin, 1926.
9. Weibull H.R.W. Pressures recorded in partially closed chambers at explosion of TNT charges. *Annals of the New York Academy of Sciences*, 1968, **152**, Article 1, 357–361.
10. Smith P.D. and Hetherington J.G. *Blast and Ballistic Loading of Structures*. Butterworth-Heinemann, 1994.

4 Structural response to blast loading

Notation

d	depth of beam
E	Young's modulus of elasticity
F	peak blast load of idealized triangular pulse
$F(t)$	blast load from idealized triangular pulse
i_r	specific reflected impulse
i_s	specific side-on impulse
$\bar{i}_s$	scaled impulse
I	impulse
K	stiffness
K'	empirical constant
K_e	equivalent stiffness
K_L	load factor
K_M	mass factor
K_{LM}	load–mass factor
K_s	stiffness factor
KE	kinetic energy
L	length of beam
m	mass of human subject
M	structure mass
M_e	equivalent mass
p_0	atmospheric pressure
p_r	peak reflected pressure
p_s	peak side-on overpressure
$\bar{p}_s$	scaled pressure
P	blast load
P_e	equivalent blast load
R	range, structure resistance, static load to cause same deflection as blast load
R_B	radius for category B damage
t	time
t_d	duration of idealized triangular blast load
t_m	time to reach maximum dynamic displacement
T	natural period of vibration of structure

$\bar{T}$	scaled blast wave duration
U	strain energy
V	dynamic reaction
W	charge mass
WD	work done by blast load
W_0	maximum displacement
x	displacement
x_{max}	maximum dynamic displacement
x_{st}	static displacement
$\bar{x}$	centroid of deflected shape
Z	scaled distance
ρ	density
σ_y	yield stress
ω	natural circular frequency of vibration

Introduction

In assessing the behaviour of a blast-loaded structure it is often the case that the calculation of final states is the principal requirement for a designer rather than a detailed knowledge of its displacement–time history. To establish the principles of this analysis, the response of a single degree of freedom (SDOF) elastic structure is considered and the link between the duration of the blast load and the natural period of vibration of the structure established. This leads to the concept of 'impulsive' and 'quasi-static' response regimes and the representation of such response on pressure–impulse diagrams. Pressure–impulse diagrams both for building structures and other targets such as personnel are described. The principles of analysis for an SDOF system are extended to specific structural elements which can then be converted back to equivalent lumped mass structures by means of load and mass factors. Total structural resistance can thus be represented by the sum of an inertial term (based on the mass of the structure) and the so-called 'resistance function' (based on the structure's geometrical and material properties) which act in opposition to the applied blast load.

Elastic SDOF structure

Consider a structure which has been idealized as a SDOF elastic structure and which is to be subjected to a blast load idealized as a triangular pulse delivering a peak force F. The positive phase duration of the blast load is t_d. The situation is illustrated in Fig. 4.1.

The load pulse is described by the equation

$$F(t) = F\left(1 - \frac{t}{t_d}\right) \tag{4.1}$$

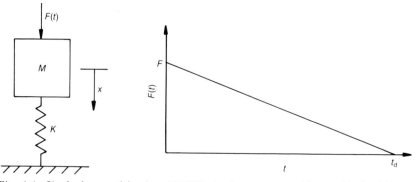

Fig. 4.1. Single degree of freedom (SDOF) elastic structure subject to idealized blast pulse [11]

This blast load will deliver an impulse I to the target structure given by the equation

$$I = \tfrac{1}{2} F t_d \qquad (4.2)$$

here I is the area beneath the load function for $0 < t < t_d$. The equation of motion for this structure is

$$M\ddot{x} + Kx = F\left(1 - \frac{t}{t_d}\right) \qquad (4.3)$$

If we confine the problem to response for times less than the positive phase duration, the solution can be written as

$$x(t) = \frac{F}{K}(1 - \cos \omega t) + \frac{F}{K t_d}\left(\frac{\sin \omega t}{\omega} - t\right) \qquad (4.4)$$

where $\omega\,[=\sqrt{(K/M)}]$ is the natural frequency of vibration of the structure.

By limiting analysis to the worst case of response the maximum dynamic structure displacement, x_{max}, is required which will occur when the velocity of the structure is zero. Differentiating eqn (4.4) and setting dx/dt to zero gives

$$0 = \omega \sin(\omega t_m) + \frac{1}{t_d}\cos(\omega t_m) - \frac{1}{t_d} \qquad (4.5)$$

In this equation t_m is the time at which the displacement reaches x_{max}. Equation (4.5) may be solved to obtain a relationship of the general form

$$\omega t_m = f(\omega t_d) \qquad (4.6)$$

From this it is clear that a similar form of equation can be obtained for maximum dynamic displacement

$$\frac{x_{max}}{F/K} = \psi(\omega t_d) = \psi'\left(\frac{t_d}{T}\right) \qquad (4.7)$$

where ψ and ψ' are functions of ωt_d and t_d/T, respectively and T is the natural period of response of the structure. Solutions of this form indicate that there is a strong relationship between T and t_d. To proceed further, consider the relative magnitudes of these quantities.

Positive phase duration and natural period

Positive phase long compared with natural period

First consider the situation where t_d is much longer than T. In the limit the load may be considered as remaining constant whilst the structure attains its maximum deflection. For example, this could be the case for a structure loaded by a blast from a domestic gas explosion. In this case the maximum displacement x_{max} is solely a function of the peak blast load F and the stiffness K. The situation can be represented graphically, and Fig. 4.2 shows the variation of both blast load and the development of structure resistance, $R(t)$, with time. The structure is seen to have reached its maximum displacement before the blast load has undergone any significant decay. Such loading is referred to as quasi-static or pressure loading.

Positive phase short compared with natural period

Now consider the situation in which t_d is much shorter than T. In this case the load has finished acting before the structure has had time to respond significantly — most deformation occurs at times greater than t_d. Thus, we can say that displacement is a function of impulse, stiffness and mass which can be represented graphically (Fig. 4.3).

Inspection of these graphs indicates that the blast load pulse has fallen to zero before any significant displacement occurs. In the limit, the blast

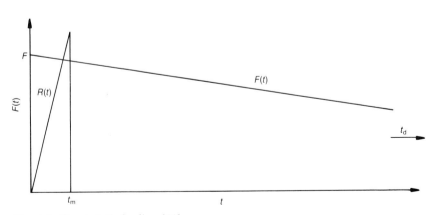

Fig. 4.2. Quasi-static loading [11]

BLAST EFFECTS ON BUILDINGS

load will be over before the structure has moved: this situation is described as impulsive loading.

Positive phase duration and natural period similar

In this case, with t_d and T approximately the same, the assessment of response is more complex, possibly requiring complete solution of the equation of motion of the structure.

These three regimes can be summarized in terms of the product ωt_d (which is proportional to the ratio of T to t_d) as indicated below

$$0 \cdot 4 > \omega t_d \quad \left[\propto \frac{t_d \text{ (short)}}{T \text{ (long)}} \right] \quad \text{Impulsive}$$

$$40 < \omega t_d \quad \left[\propto \frac{t_d \text{ (long)}}{T \text{ (short)}} \right] \quad \text{Quasi-static}$$

$$0 \cdot 4 < \omega t_d < 40 \quad \left[\frac{t_d}{T} \approx 1 \right] \quad \text{Dynamic.} \tag{4.8}$$

Evaluation of the limits of response

In the case of quasi-static loading the load pulse can be idealized as shown in Fig. 4.4(a) while the elastic structure resistance can be represented by the graph of Fig. 4.4(b).

The basic principle of the analysis is to equate the work done on the structure and the strain energy acquired by the structure as it deforms. The work done by the load as it causes a displacement x_{max} is WD given by

$$WD = Fx_{max} \tag{4.9}$$

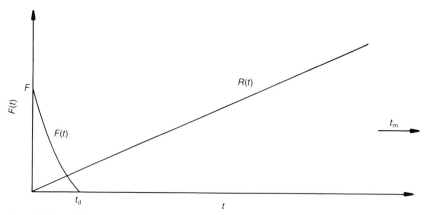

Fig. 4.3. Impulsive loading [11]

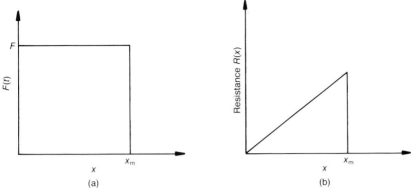

Fig. 4.4. Idealized load and resistance–deflection function for quasi-static loading [11]

The strain energy acquired by the structure, U, is the area beneath the resistance displacement graph given by

$$U = \tfrac{1}{2} K x_{max}^2. \tag{4.10}$$

Equating WD and U (after some rearrangement) results in

$$\frac{x_{max}}{(F/K)} = \frac{x_{max}}{x_{st}} = 2 \tag{4.11}$$

where x_{st} is the static displacement that would result if the force F were applied statically. Equation (4.11) represents the so-called dynamic load factor (DLF) which gives the upper bound of response and is called the 'quasi-static asymptote'.

When an impulse is delivered to a structure it produces an instantaneous velocity change: momentum is acquired and the structure gains kinetic energy which is converted to strain energy. The impulse causes an initially stationary structure to acquire a velocity $\dot{x}_0\,(= I/M)$. From this the kinetic energy delivered, KE, is given by

$$KE = \frac{1}{2} M \dot{x}_0^2 = \frac{I^2}{2M} \tag{4.12}$$

The structure will acquire the same strain energy U as before because it displaces by x_{max}. Thus if KE and U are equated, after some rearrangement we obtain

$$\frac{x_{max}}{F/K} = \frac{x_{max}}{x_{st}} = \frac{1}{2} \omega t_d \tag{4.13}$$

which is the equation of the 'impulsive asymptote' of response.

If these two asymptotes are drawn on a response curve of ωt_d against

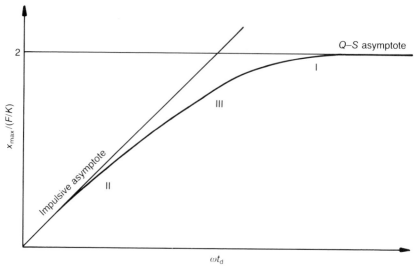

Fig. 4.5. Graphical representation of quasi-static (I), impulsive (II) and dynamic (III) response [11]

$x_{max}/(F/K)$, the actual response of the structure can then be sketched without recourse to further analysis as shown in Fig. 4.5.

The three regimes of quasi-static, impulsive and dynamic response are identified on the resulting graph as regions I, II and III, respectively.

Pressure–impulse diagrams

It is now easily possible to convert Fig. 4.5 to a pressure–impulse (P–I) diagram which allows the load–impulse combination that will cause a specified level of damage to be assessed very readily. The bounds on behaviour of a target structure are characterized by a pressure (or, as here, force) and a total impulse I (as here) or a specific impulse i_s or i_r.

Equation (4.11) can be rewritten as

$$\frac{2F}{Kx_{max}} \left[\propto \frac{\text{maximum load}}{\text{maximum resistance}} \right] = 1 \qquad (4.14)$$

which is the equation of a modified quasi-static asymptote plotted on a graph with ordinate $2F/Kx_{max}$. If the abscissa of Fig. 4.5 is multiplied by the inverse of the ordinate of Fig. 4.5, and recalling that $\omega = \sqrt{(K/M)}$ we obtain

$$(Ft_d/Kx_{max})\sqrt{K/M} = 2I/(x_{max}\sqrt{KM}) \qquad (4.15)$$

which is a non-dimensionalized impulse. Noting that the impulsive asymptote of Fig. 4.5 is given by eqn (4.13), using an abscissa $I/(x_{max}\sqrt{KM})$

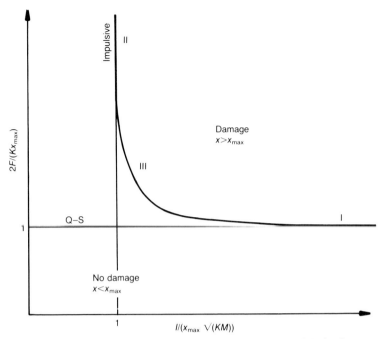

Fig. 4.6. Non-dimensionalized pressure–impulse diagram for SDOF elastic system [2]

means that the new impulsive asymptote is given by $I/(x_{max}\sqrt{KM}) = 1$ and the response curve can be re-plotted to obtain Fig. 4.6.

The form of this graph allows easy assessment of response to a specified load. Once a maximum displacement or damage level is defined this curve then indicates the combinations of load and impulse that will cause failure. Combinations of pressure and impulse that fall to the left of and below the curve will not induce failure while those to the right and above the graph will produce damage in excess of the allowable limit.

Whereas the foregoing relates to analytically derived pressure–impulse diagrams it is possible to derive P–I curves from experimental evidence or real-life events. P–I curves have been derived from a study of houses damaged by bombs dropped on the UK in the Second World War [1]. The results of such investigations are used in the evaluation of safe stand-off distances for explosive testing in the UK. In this instance the axes of the curves are simply side-on peak overpressure p_s and side-on specific impulse i_s as shown in Fig. 4.7. These curves can also be used with reasonable confidence to predict the damage to other structures such as small office buildings and light-framed factories.

The levels of damage are less precise than those obtained by analysis. As quoted in [1], A corresponds to almost complete demolition of the building and B means damage severe enough to necessitate demolition.

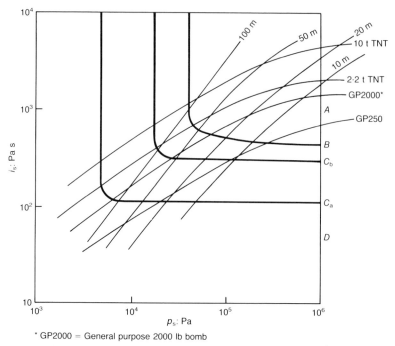

*GP2000 = General purpose 2000 lb bomb

Fig. 4.7. Isodamage curves with range–charge weight overlay for brick-built houses [11]

Category C_b implies damage rendering the house temporarily uninhabitable with the roof and one or two external walls partially collapsed. Load-bearing partitions would be severely damaged and would require replacement. In contrast, category C_a indicates relatively minor structural damage though still sufficient to make the house temporarily uninhabitable with partitions and joinery being wrenched from fixings. Finally, category D refers to damage calling for urgent repair but is not severe enough to make the building uninhabitable: there would be damage to ceilings and tiling and more than 10% of glazing would be broken.

The use of pressure–impulse diagrams in conjunction with blast parameter versus scaled distance graphs allows the development of equations to describe specific damage levels. An example for the curves above is of the general form

$$R = \frac{K'W^{1/3}}{[1+(^{3175}/W)^2]^{1/6}} \qquad (4.16)$$

where R is range in metres, W is mass of explosive in kilograms of TNT and K' is an empirical constant. The value of K' in the equation above

STRUCTURAL RESPONSE TO BLAST LOADING

Table 4.1. Overpressure and scaled distance for various types of blast damage

Structural element	Failure mode	1 t TNT		10 t TNT	
		p_s (kPa)	Z (m/kg$^{1/3}$)	p_s (kPa)	Z (m/kg$^{1/3}$)
Window panes	5% broken	1·1	72·2	0·7	96·0
	10% broken	2·5	38·6	1·7	51·6
	90% broken	6·3	19·6	4·2	26·9
Houses	Tiles displaced	4·5	25·6	2·9	34·7
	Doors/window frames blown in	9·1	14·6	6·0	20·4
	Category D	5	23·7	3·1	33·6
	Category C_a	13	11·4	8	16·1
	Category C_b	28	6·5	17	9·2
	Category B	80	3·6	36	5·6
	Category A	185	2·4	80	3·6

giving the radius of B type damage (R_B) is 5·6. Radii for types A and C_b are given approximately by $0·675\,R_B$ and $1·74\,R_B$, respectively.

Further analysis allows the addition of 'range–charge weight' overlays to P–I curves enabling the damaging potential of a particular threat defined in terms of explosive yield and stand-off to be assessed. Such overlays are included in Fig. 4.7 for a number of threats and distances.

As a complement to the information presented in Fig. 4.7, Table 4.1 (extracted from [3]) gives the overpressure and scaled distance for a number of types of building blast damage from charges of TNT of 1 and 10 tonnes.

Table 4.1 clearly demonstrates the importance of positive phase duration as well as overpressure in determining damage: a larger charge can cause the same level of damage as a smaller charge even though the associated overpressure is less. This is because the larger device generates a pulse of longer duration. Thus impulse should be given equal consideration with overpressure when assessing the damage potential of a given threat.

Pressure–impulse diagrams for human response to blast loading

Generally, three categories of blast-induced injury are identified. These are:

(a) *Primary injury:* due directly to blast wave overpressure and duration which can be combined to form specific impulse. Overpressures are induced in the body following arrival of the blast and the level of injury sustained depends on a person's size, gender and (possibly) age. The location of most severe injuries is where density differences between adjacent body tissues are greatest. Likely damage sites thus include the lungs which are prone to haemorrhage and oedema

(collection of fluid), the ears (particularly the middle ear) which can rupture, the larynx, trachea and the abdominal cavity.
(b) *Secondary injury:* due to impact by missiles (e.g. fragments from a weapon's casing). Such missiles produce lacerations, penetration and blunt trauma (a severe form of bruising).
(c) *Tertiary injury:* due to displacement of the entire body which will inevitably be followed by high decelerative impact loading when most damage occurs. Even when the person is wearing a good protective system, skull fracture is possible.

In the particular case of primary injury associated with damage to the lungs, P–I diagrams have been developed as shown in Fig. 4.8(a) constructed by Baker et al. [2] from a number of sources.

The axes are scaled pressure $\bar{p}_s = p_s/p_0$, where p_s is peak incident pressure and p_0 is ambient (usually atmospheric) pressure and scaled impulse $\bar{i}_s$. This is obtained by defining a scaled positive phase duration $\bar{T} (= t_d p_0^{1/2}/m^{1/3})$, where m is the mass of the person and the blast wave is taken as a triangular pulse. Then

$$\bar{i}_s = \tfrac{1}{2}\bar{p}_s \bar{T} = i_s/p_0^{1/2} m^{1/3} \tag{4.17}$$

The addition of a range–charge weight overlay to Fig. 4.8(a) leads, with a little modification, to the survival prediction curves of Fig. 4.8(b) which relate to a 70 kg man [3].

Similar diagrams are available for assessment of primary injury to the ear, although since less information is available about variability of response as blast duration changes, damage levels are generally related to an overpressure value as summarized in Table 4.2 [3].

It should be noted that secondary and tertiary injuries can be just as likely and possibly more severe than primary effects. For secondary injury criteria reference should be made to Ahlers [4] summarized in [5, 11]. In the case of tertiary injuries [6], summarized in [4, 5, and 11] is recommended.

Table 4.2. *Summary of overpressure values*

Z (m/kg$^{1/3}$)	p_s (kPa)	Eardrum damage: %
5·63	35·6	5
4·88	45·4	10
3·93	67·7	25
3·13	105·0	50
2·54	163·0	75
2·14	243·0	90

STRUCTURAL RESPONSE TO BLAST LOADING

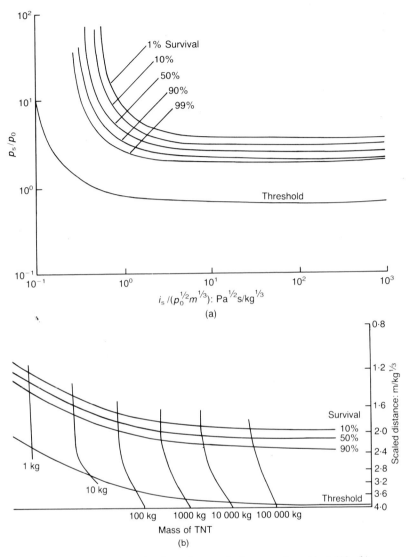

Fig. 4.8. (a) Isodamage curves for lung damage to humans from blast [2]. (b) Isodamage curves with modified range–charge weight overlay for lung damage to 70 kg man [3]

Energy solutions for specific structural components
Elastic analysis

The approach adopted here is essentially the Rayleigh–Ritz method of analysis [7] and is suitable for specific structural members in an uncoupled analysis. This means that the response of structural elements to a blast load may be considered in isolation assuming that the support conditions are essentially rigid. In the implementation described here the approach produces worst cases of response rather than displacement-time histories.

Firstly, a mathematical representation of the deformed shape is selected for the structure which satisfies all the necessary boundary conditions relating to displacement. Then, by operating on the deformed shape, the curvature and hence strain of deformation is obtained from which the total strain energy of the element can be calculated.

Consideration must now be given to the nature of the blast load impinging on the structure. If the loading is impulsive then a calculation of total kinetic energy delivered to the structure is made. If, however, the load is quasi-static, the work done by the load is found by considering the work done on a small element of the structure then integrating over the loaded area. In the impulsive realm, response is evaluated by equating the kinetic energy acquired to the strain energy produced in the structure. In the quasi-static realm response is assessed by equating the work done by the load to strain energy. Having done this, it is possible to quantify particular aspects of response such as maximum displacement, maximum strains and maximum stresses.

If this technique is used, for example, to analyse the response of a cantilever of depth d to impulsive loading then the maximum displacement W_0 of the structure of length L made of material of Young's modulus E and density ρ loaded by a reflected specific impulse i_r is given by

$$\frac{W_0}{L} = C_1 \left[\frac{L}{d}\right]\left[\frac{i_r}{d\sqrt{(E\rho)}}\right] \qquad (4.18)$$

where the exact value of the coefficient C_1 (equal to approximately two) will depend on the choice of deflected shape of the structure.

If the same cantilever is now loaded quasi-statically and an analysis is carried out assuming that the blast pressure on the structure remains steady during structure deformation (i.e. the peak reflected overpressure, p_r, remains constant), the resulting maximum displacement is now obtained as

$$\frac{W_0}{L} = C_2 \left[\frac{L}{d}\right]^3 \left[\frac{p_r}{E}\right] \qquad (4.19)$$

where the exact value of C_2 (equal to approximately three) depends on the deflected shape chosen.

Plastic analysis

The approach to the analysis of structural elements that deform beyond their elastic limits proceeds in a similar way. For example, consider a simply supported beam loaded impulsively made of material with stress–strain characteristics idealized as 'rigid–plastic'.

The analytical approach described above leads to an expression for maximum (central) displacement W_0 given by

$$W_0/L = i_r^2 L/4\sigma_y \rho d^3 \qquad (4.20)$$

where σ_y is the yield strength of the material.

Just as in the case of the SDOF system detailed above, P–I diagrams can be constructed for these or any other structural elements.

Lumped mass equivalent SDOF systems

The analysis presented above is valuable, though for more complex structural elements and load configurations, implementation of this approach could be rather time-consuming. To aid assessment of response, the behaviour of complex structures can be approached by representing the structure as a SDOF lumped-mass system — the so-called equivalent system. The equation of motion so derived will be similar to eqn (4.3) and can be solved either analytically or numerically to obtain a deformation–time history for the structure. Often, as described above, maximum displacement may be all that is required. This approach, although failing to provide detailed aspects of response, allows a good insight into important features of behaviour and will, furthermore, give an over-assessment of response. This conservative analysis, therefore, has attraction for the designer.

Equation of motion for an SDOF system

Equation (4.3) can be rewritten with the spring resistance term replaced by a more general resistance function $R(x)$ as

$$M\ddot{x} + R(x) = P(t) \qquad (4.21)$$

In creating an equivalent SDOF structure it must be realized that real structures are multi-degree of freedom systems where every mass particle has its own equation of motion. Thus, to simplify the situation it is necessary to make assumptions about response and in particular characterize deformation in terms of a single point displacement. Examples are shown in Fig. 4.9(a) and (b) where in the SDOF system the suffix e means equivalent.

The method relies, as in the examples above, on considering the energies of the real structure and the equivalent system and equating them. This means that, by ensuring equal displacements and velocities in the two systems, kinematic similarity is maintained. The complete energy relationship may be written as

$$WD = U + KE \qquad (4.22)$$

BLAST EFFECTS ON BUILDINGS

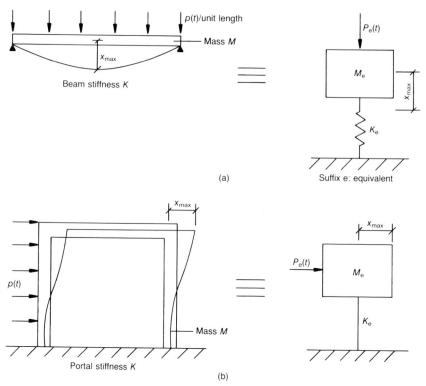

Fig. 4.9. Real and equivalent structural systems [11]

In the derivations of impulsive and quasi-static response described above it was this equation that was simplified in analysing the two extremes of behaviour.

Example of the approach

Consider a simply supported beam responding elastically as shown in Fig. 4.9(a). Firstly, a suitable deformed shape is assumed such as the displacement under a uniformly distributed static load. The evaluation of work done, strain energy and kinetic energy for the beam is then made. The equivalent system will be as shown in Fig. 4.9(a) which will have the same maximum displacement W_0 and maximum initial velocity $\dot{W}_0$ as for the real structure. The evaluation of WD, U and KE is much simpler here

$$WD = P_e(t)W_0$$
$$U = \tfrac{1}{2}K_e W_0^2$$
$$KE = \tfrac{1}{2}M_e \dot{W}_0^2 \qquad (4.23)$$

Equating the work terms from the two analyses leads to the definition of a load factor, K_L as

$$K_L = P_e(t)/P(t) = 16/25 = 0.64 \qquad (4.24)$$

with the value particular to the beam given.

Equating strain energy for the two systems and recalling that the stiffness of a flexing structure is given by load per unit displacement, definition of a stiffness factor, K_s is given by

$$K_s = K_e/K = 0.6401 \qquad (4.25)$$

with the value for this particular structural element as shown.

Finally, equating kinetic energy for the two systems leads to definition of a mass factor, K_M, as

$$K_M = M_e/M = 0.504 \qquad (4.26)$$

with the particular value for this structure as shown.

It is worth noting that the load and stiffness factors are very similar and it is general practice to set them equal: thus we have $K_s = K_L$. Also it is often convenient to define a load–mass factor $K_{LM} = K_M/K_L$ the value of which approach is easily seen. The equivalent system equation of motion can be expressed in terms of mass and load factors. Division by load factor (assuming that stiffness and load factors are equal) yields, for an elastic system

$$K_{LM}M\ddot{X} + Kx = P(t) \qquad (4.27)$$

Thus, an equivalent system equation of motion can be derived merely by factoring the actual structure mass by K_{LM}.

One further important calculation that should be made is of the dynamic support reactions generated by the blast load. For the specific example of the simply supported beam, by considering half of the beam, moments are taken about the centre of inertial resistance of the beam which is taken to coincide with the centroid of the shape swept out by the beam as it deforms. For the shape chosen here the centroid can be shown to be $\bar{x} = 61L/192$ from the end of the beam. Taking moments about this point leads to an equation for dynamic reaction $V(t)$ as

$$V(t) = 0.393R + 0.107P(t) \qquad (4.28)$$

where R is the static load needed to cause the same deflection as the blast load [$P(t)$]. The same analysis can be used to develop load and mass factors for plastically deforming structures.

Appendix B gives load and mass factors as well as dynamic reactions for a variety of one-way spanning structural elements for both elastic and plastic response. The reader is also referred to the text by Biggs [8] in which the concepts outlined above were originally developed.

Resistance functions for specific structural forms

As noted earlier in this chapter, the term $R(x)$ is known as the resistance function of the element. In developing the resistance function for a structural element, it is assumed that the element will offer essentially the same resistance to deflection when deformed dynamically as it will when deformed quasi-statically. The only adjustment incorporated is an enhancement in the ultimate resistance of the element, due to the improvement in strength observed in dynamic loading. The general form of the resistance function for a common structural form will be presented below.

Resistance functions — general description

The resistance–deflection function for a structural element is strictly a graph of the uniform pressure which would be necessary to cause deflection at the central point of the element during its transient displacement.

A typical resistance–deflection curve for a laterally restrained element is shown in Fig. 4.10. The initial portion of the curve is due primarily to flexural action. If lateral restraint prevents small motions, in-plane compressive membrane forces are developed. However, the increased capacity due to these forces is normally neglected [9] and is not shown in the figure. Following the loss of flexural capacity, the provision of adequate lateral restraint may permit the development of tensile membrane action. The increase in resistance with increasing deflection up to incipient failure is shown as the dashed line in the figure.

In practice, when designing for smaller deflections, a simplified function

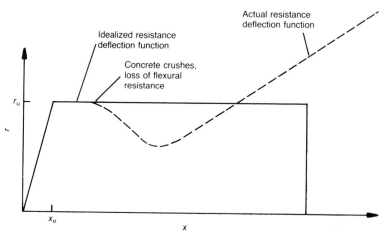

Fig. 4.10. *Typical resistance–deflection curve for large deflections*

STRUCTURAL RESPONSE TO BLAST LOADING

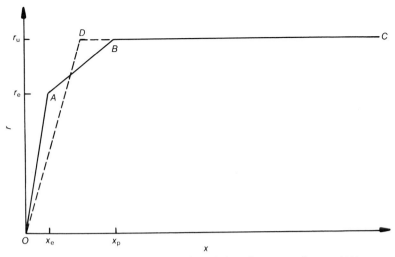

Fig. 4.11. Simplified resistance function for reinforced concrete element [11]

is used which is a prediction of the resistance which the element would offer in a quasi-static test. Figure 4.11 shows the resistance function which would be adopted for a reinforced concrete wall with built-in supports along two edges but no support along the other two, i.e. a one-way-spanning slab. The initial part of the graph, OA, represents the elastic deformation of the slab. At the point A, yield lines develop along the built-in supports, allowing rotation. During this phase, AB, deformations in the central part of the slab are elastic whilst rotation occurs in the yield lines. The increments of displacement during this elasto-plastic phase are estimated from the elastic stiffness of a simply supported one-way spanning slab. More complicated structural forms, e.g. a two-way-spanning slab with built in supports, may have two or more elasto-plastic phases as the yield-line system develops progressively in the slab. The phase BC represents the ultimate resistance of the element once a sufficient system of yield lines has been established to form a collapse mechanism in the slab. This fully plastic phase will last until failure at one of the supports occurs. For the case of two-way spanning slabs it is possible for failure to occur along two opposite supports whilst the other pair of supports remains intact. In this case the slab will continue to offer resistance, but at a lower level. Experience shows that a negligible error is introduced by replacing the resistance function OABC with the bilinear function ODC, provided the area under both graphs is the same. This approach is adopted in the document TM5-1300 [9]. Tables for deriving the components of the resistance function for a variety of one-way spanning elements may be found in Appendix B.

Conclusions

Methods for analysing the response of blast-loaded structures have been presented. It is worth noting that many of the concepts of providing protection are by no means new. The work by Christopherson [10] presented the essence of the approach which has been the basis of a number of subsequent publications. Of particular importance to the designer is the use of these principles in establishing the factors that allow formation of equivalent SDOF systems which, when combined with the loading information presented in Chapter 3, form the basis of the subsequent design process. For structures which respond within the impulsive regime, design is normally based upon an energy solution in which the kinetic energy delivered [eqn (4.12)] is equated to the strain energy produced in deforming to a limited deflection — that is the area under the resistance function. In the quasi-static regime it is the work done on the structure by the load that must be equated to the strain energy produced. For response within the intermediate, or dynamic regime, the response charts presented in Appendix C, which are based on numerical solutions of the full equation of motion, may be used in the design process. Further details about analysis of structures subjected to blast loading can be obtained from [11]. The application of these techniques to the design of elements in reinforced concrete and structural steel is more fully described in Chapter 5.

References

1. Jarrett D.E. Derivation of British explosives safety distances. *Annals of the New York Academy of Sciences*, 1968, **152**, Article 1, 18–35.
2. Baker W.E., Cox P.A., Westine P.S., Kulesz J.J. and Strehlow R.A. *Explosion Hazards and Evaluation*. Elsevier, London, 1983.
3. Merrifield R. *Simplified calculations of blast induced injuries and damage*. Health and Safety Executive Specialist Inspector, April 1993, Report No. 37.
4. Ahlers E.B. Fragment hazard study. *Minutes of the 11th explosives safety seminar, Vol. 1*. Armed Services Explosives Safety Board, Washington DC, 1969.
5. Baker W.E., Westine P.S., Kulesz J.J., Wilbeck J.S. and Cox P.A. A manual for the prediction of blast and fragment loading on structures. Department of Energy, Amarillo, Texas, 1980, DOE/TIC-11268 US.
6. Baker W.E., Kulesz J.J., Ricker R.E., Bessey R.L., Westine P.S., Parr V.B. and Oldham G.A. *Workbook for predicting pressure wave and fragment effects of exploding propellant tanks and gas storage vessels*. NASA Lewis Research Centre, 1975 (reprinted 1977), NASA CR-134906.
7. Todd J.D. *Structural theory and analysis*. MacMillan, 1974.
8. Biggs J.M. *Introduction to structural dynamics*. McGraw-Hill, New York, 1964.
9. US Department of the Army Technical Manual, TM5-1300. *Design of structures to resist the effects of accidental explosions*. Washington DC, 1990.

10. Christopherson D.G. *Structural defence*. Ministry of Home Security Research and Experiments Department, January 1946, Report RC 450.
11. Smith P.D. and Hetherington J.G. *Blast and ballistic loading of structures*. Butterworth–Heinemann, 1994.

5 Design of elements in reinforced concrete and structural steel

Notation

A_d	area of diagonal bars at the support within a width b
A_s	area of tension reinforcement within a width b
A_s'	area of compression reinforcement within a width b
A_v	total area of stirrups in tension within a distance s and a width b
A_w	area of web of steel section
b	width of flexural member
d	distance from extreme compression fibre to centroid of tension reinforcement
d'	distance from extreme compression fibre to centroid of compression reinforcement
d_c	distance between the centroids of the compression and tension reinforcement
DIF	dynamic increase factor
E	modulus of elasticity
E_c	modulus of elasticity of concrete
E_s	modulus of elasticity of reinforcement or steel section
f	stress
f_{cu}	static ultimate compressive strength of concrete at 28 days
f_{dcu}	dynamic ultimate compressive strength of concrete at 28 days
f_{dc}	dynamic design stress for concrete
f_{ds}	dynamic design stress for reinforcement or steel section
f_{du}	dynamic ultimate stress of reinforcement or steel section
f_{dv}	dynamic design shear stress for steel section
f_{dy}	dynamic yield stress of reinforcement or steel section
f_u	static ultimate stress of reinforcement
f_y	static yield stress of reinforcement
F	coefficient for moment of inertia of cracked section
H	span height
i	unit positive impulse
i_r	unit positive normal reflected impulse
I	moment of inertia
I_c	moment of inertia of cracked concrete section of width b
K_e	elastic unit stiffness

DESIGN OF ELEMENTS IN REINFORCED CONCRETE AND STRUCTURAL STEEL

K_{ep}	elasto-plastic unit stiffness
K_E	equivalent elastic unit stiffness
K_L	load factor
K_{LM}	load–mass factor
K_M	mass factor
L	span length
m	unit mass
m_e	effective unit mass
M_n	ultimate negative moment capacity
M_p	ultimate positive moment capacity
n	modular ratio
p	pressure
p_{max}	peak pressure
p_r	peak positive normal reflected pressure
P	load
r_e	elastic unit resistance
r_{ep}	elasto-plastic resistance
r_u	ultimate unit resistance
R	total internal resistance
R_u	total ultimate resistance
s	spacing of stirrups in the direction parallel to the longitudinal reinforcement
S	plastic section modulus
t	time
t_d	duration of positive phase of blast pressure
t_m	time at which maximum deflection occurs
T_c	thickness of concrete section
T	effective natural period of vibration
v_c	ultimate shear stress permitted in an unreinforced concrete web
v_u	ultimate shear stress
V	support reaction
V_d	ultimate direct shear capacity of the concrete of width b
V_p	ultimate shear capacity of steel section
V_s	shear at the support per unit width
x	depth of neutral axis below extreme compression fibre in a flexural member
X	deflection
X_m	maximum transient deflection
X_E	equivalent elastic deflection
Z	elastic section modulus
α	coefficient applied to total internal resistance R to determine dynamic reaction V
β	coefficient applied to total load P to determine dynamic reaction V
γ_m	partial safety factor for material
δ	side-sway deflection
ϵ	strain
θ	support rotation angle
μ	ductility ratio

ρ density
ρ_s tension reinforcement ratio
ρ_s' compression reinforcement ratio

Objectives

The prime objective in the design of blast load-resisting structural elements is to provide sufficient ductility to enable the element to deflect by an amount consistent with the degree of damage permitted; this will entail an initial design based upon extensive flexural plastic deformation. In so deforming, the element should not fail prematurely due to other load effects, for example shear or local instability.

Unless the element is to be subjected to repeated blast loading, for example in a test facility, the design should be based on the ultimate limit state. Simple supports should be avoided wherever possible and joints between elements should be carefully detailed to facilitate load transfer (see Plates 13, 14).

The overall design process is summarized in the logic flow chart presented in Appendix D.

Plate 13. Failure of the connections in a precast concrete frame

Plate 14. Failure due to insufficient ties, poor concrete quality and detailing of joints

Design loads

The blast loading for which resistance is to be provided is likely to be an extreme event, and as such has a low probability of occurrence. Thus, it is appropriate to set the partial safety factor for blast loading to a value of unity.

Dead loads, storage and other permanent loads should also be assigned a partial safety factor of unity.

The magnitude of imposed loads and wind loads acting simultaneously with the blast load are likely to be only a small proportion of their respective design values. A factor of 0·33 may be applied to the design values of the variable imposed and wind loads when acting in combination with the blast load, as recommended in BS 5950: Part 1 [1] and BS 8110: Part 1 [2].

When other loads are present at the time of the blast loading they may be assumed to act constantly throughout the application of the blast load. The effect of these other loads generally will be to reduce the effective resistance of an element. However, where mass is associated with such loads there may be a beneficial effect as a result of the inertial effects of these loads.

Design strengths

The design should generally be based upon the characteristic strength of materials — that is with a partial safety factor of unity, unless there is evidence to show that the mean strengths of a particular material are generally higher than the specified minimum. For example, the characteristic yield stress for grade 50 or lower structural steel may be increased by 10% in design calculations involving blast loading.

Under the action of rapidly applied loads the rate of strain application increases and this may have a marked influence on the mechanical properties of structural materials. In comparison with the mechanical properties under static loading the effects may be summarized as follows:

(a) The yield stress of structural steel or steel reinforcement bars, f_y, increases significantly to the dynamic yield stress, f_{dy}.
(b) The ultimate tensile strength of structural steel or steel reinforcement bars, f_u, in which account is taken of strain hardening effects, increases slightly to the dynamic ultimate strength, f_{du}.
(c) The compressive strength of concrete, f_{cu}, increases significantly to the dynamic compressive strength, f_{dcu}.
(d) The modulus of elasticity of both steel and concrete remains insensitive to the rate of loading.
(e) The elongation at failure of structural steel is relatively insensitive to the rate of loading.

The factor by which the static stress is enhanced in order to calculate the dynamic stress is known as the dynamic increase factor (DIF). Typical values of DIF for structural steel and reinforced concrete are given in Table 5.1. The dynamic stress to be used in the design of reinforced concrete

Table 5.1. *Dynamic increase factors (DIF) for design of reinforced concrete and structural steel elements*

Type of stress	Concrete	Reinforcing bars		Structural steel	
	f_{dcu}/f_{cu}	f_{dy}/f_y	f_{du}/f_u	f_{dy}/f_y*	f_{du}/f_u
Bending	1·25	1·20	1·05	1·20	1·05
Shear	1·00	1·10	1·00	1·20	1·05
Compression	1·15	1·10	—	1·10	—

* Minimum specified f_y for grade 50 steel or less may be enhanced by the average strength increase factor of 1·10.

and structural steel elements depends upon the deformation or damage limits imposed — these are considered later on in the chapter.

Deformation limits

The controlling criterion in the design of blast-resistant structural elements is normally a limit on the deformation or deflection of the element. In this way the degree of damage sustained by the element may be controlled. The damage level that may be tolerated in any particular situation will depend on what is to be protected, for example, the structure itself, the occupants of a building or equipment within the building.

There are two methods by which limiting element deformations may be specified: by using the support rotation, θ (see Fig. 5.1) and the ductility ratio

$$\mu = \frac{\text{total deflection}}{\text{deflection at elastic limit}} = \frac{X_m}{X_E}$$

In general, deformations in reinforced concrete elements are expressed in terms of support rotations whilst ductility ratios are used for structural steel elements.

For the protection of personnel and equipment through the attenuation of blast pressures and to shield them from the effects of primary and secondary fragments and falling portions of the structure, recommended deformation limits are given under protection category 1 in Table 5.2.

For the protection of structural elements themselves from collapse under the action of blast loading, the recommended deformation limits are given under protection category 2 in Table 5.2. It should be noted that these limits imply extensive plastic deformation of the elements and the need for subsequent repair or replacement before they may be re-used. For situations where re-use is required without repair, deformations should be maintained within the elastic range, i.e. $\mu \leq 1$. This latter design condition is likely to lead to massive and consequently costly construction.

In addition to these considerations for individual elements there remains, of course, a requirement for the overall structure to remain stable in the event of being subject to blast loading. TM5-1300 [3] recommends that the maximum member end rotation, θ, as shown in Fig. 5.2, should

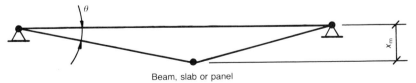

Fig. 5.1. Member support rotations [3]

Table 5.2. Deformation limits

	Protection category			
	1		2	
	θ	μ	θ	μ
Reinforced concrete beams and slabs	2°*	Not applicable	4°†	Not applicable
Structural steel beams and plates‡	2°	10	12°	20

* Shear reinforcement in the form of open or closed 'blast links' must be provided in slabs for $\theta > 1°$. Closed links (shape code 74 to BS 4466) [4] must be provided in all beams.
† Support rotations of up to 8° may be permitted when the element has sufficient lateral restraint to develop tensile membrane action. Further guidance regarding the tensile membrane capacity of reinforced concrete slabs may be found in TM5-1300 [3].
‡ Adequate bracing must be provided to assure the corresponding level of ductile behaviour.

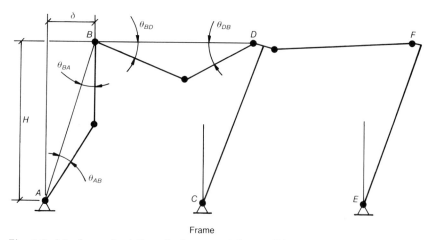

Fig. 5.2. Member end rotations for beams and frames [3]

be 2° and the maximum side-sway deflection, δ, be limited to 1/25 of the storey height, H, in framed steel structures.

Introduction to the behaviour of reinforced concrete and structural steelwork subject to blast loading
Structural response of reinforced concrete

When a reinforced concrete element is dynamically loaded, the element deflects until such time as the strain energy of the element is developed

DESIGN OF ELEMENTS IN REINFORCED CONCRETE AND STRUCTURAL STEEL

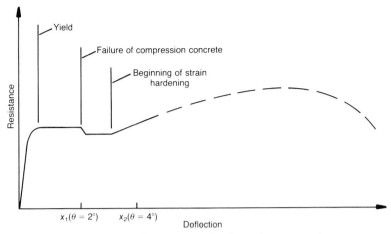

Fig. 5.3. *Typical resistance–deflection curve for flexural response of concrete elements* [3]

sufficiently to balance the energy delivered by the blast load and the element comes to rest, or fragmentation of the concrete occurs. The resistance–deflection curve shown in Fig. 5.3 demonstrates the flexural action of a reinforced concrete element.

When the element is first loaded, the resistance increases linearly with deflection until yielding of the reinforcement occurs. Thereafter the resistance remains constant with increasing deflection until, at a deflection x_1 corresponding to a support rotation, θ, of 2°, the concrete crushes in compression. Thus, for θ in the range 0°–2° the concrete is effective in resisting moment and the concrete cover on both surfaces of the element remains intact. This is referred to as a type 1 section (see Fig. 5.4a). Type 1 elements may be singly or doubly reinforced, although to cater for rebound effects some compression reinforcement is usually desirable. The ultimate moment capacity, M_p, of type 1 sections may be determined using conventional plastic theory for reinforced concrete based upon the dynamic design stresses of the concrete (f_{dc}) and the reinforcement steel (f_{ds}).

For $\theta > 2°$ the compression forces are transferred from the concrete to the compression reinforcement which results in a slight loss of capacity as shown in Fig. 5.3. In the absence of any compression reinforcement the crushing of the concrete would result in failure of the element. Sufficient compression reinforcement must be available to fully develop the tension steel, i.e. symmetrical reinforcement must be provided. The requirement for the provision of 'blast links' for support rotations in excess of 1° is to properly tie this flexural reinforcement. Elements which sustain crushing of the concrete without any disengagement of the cover on the

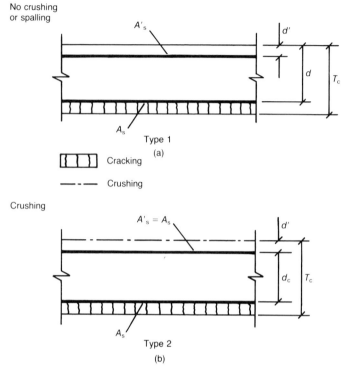

Fig. 5.4. Typical reinforced concrete cross sections [3]

Plate 15. Ductility and spalling of reinforced concrete. Photograph courtesy of Francis Walley

tensile face are known as type 2 sections (see Fig. 5.4b) and occur for θ in the range 2°–5°. The ultimate moment capacity, M_p, of type 2 sections of width b is given by

$$M_p = A_s \frac{f_{ds}}{b} d_c$$

where A_s is the area of reinforcement on each face and d_c is the distance between the centroid of the reinforcement on the tension and compression faces.

As the element is further deflected, the reinforcement enters its strain hardening region and the resistance increases with increasing deflection (Plate 15). In the absence of any tensile membrane action the blast links will restrain the compression reinforcement from buckling for a short time into its strain hardening region. At a deflection x_2 corresponding to a θ value of about 4° the element will lose its structural integrity and fail unless other forms of restraint, for example, lacing reinforcement, are used. Laced reinforced elements may be used in specialized explosive storage facilities but are unlikely to be appropriate for buildings because of the complexity of the reinforcement detailing.

Although there is little, if any, evidence of shear failures under blast loading, premature shear failures must be avoided in order to fully develop the flexural capacity of an element. The shear capacity of the concrete alone may be enhanced by providing additional shear reinforcement.

Structural response of steelwork

Structural steel can generally be considered as exhibiting a linear stress–strain relationship up to the yield point, beyond which it can strain substantially without appreciable increase in stress. This yield plateau extends to a ductility ratio, μ, of between 10 and 15. Beyond this range strain hardening occurs and after reaching a maximum stress — known as the tensile strength — a drop in stress accompanies further elongation and precedes fracture at a strain of approximately 20–30%. Typical static and dynamic stress–strain curves for steel are shown in Fig. 5.5.

Structural steels of strengths higher than grade 50 generally exhibit smaller elongations at rupture and should be used with caution when very large ductilities are a prerequisite of design.

The design plastic moment, M_p, for steel elements with $\mu \leq 3$ is given by

$$M_p = f_{ds}(Z+S)/2$$

where Z and S are the elastic and plastic section moduli, respectively. For $\mu > 3$

$$M_p = f_{ds} S$$

BLAST EFFECTS ON BUILDINGS

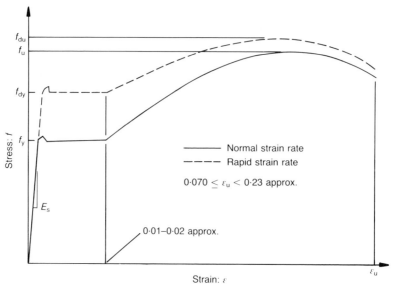

Fig. 5.5. Typical stress-strain curves for steel [3]

As with reinforced concrete sections, sufficient shear capacity must be provided and local buckling failures avoided in order to develop fully the flexural capacity of an element.

Comparison of response

Reinforced concrete is relatively massive and, as such, is more appropriate than steel to resist the close-in effects of large explosions in the impulsive regime. Steelwork is better suited to resist relatively low pressures of a quasi-static nature (Plate 16). The massive nature of reinforced concrete implies 'stocky' sections whose ultimate capacity is reasonably predictable. The slender nature of structural steel sections can cause local instability and unpredictable ultimate capacities (Plate 17).

There are two other significant differences between the materials. First, the rebound in concrete structures is small because cracking causes internal damping. In steel, the rebound can be quite large, particularly for short duration loads on relatively flexible elements. Therefore, steel structures must be designed to support significant reversals of loading. Second, in reinforced concrete, separate reinforcing steel is provided to resist flexure, shear and torsion. In steel, complex stress combinations occur which are difficult to predict and which can potentially cause distress. Stress concentrations at welds and notches must also be carefully considered if the full strength of the section is to be realized.

Introduction to the design of reinforced concrete elements to resist blast loading

The design methods recommended in this chapter are based on those described in [3]. As such, the design techniques set forth are based upon the results of numerous full- and small-scale structural response and explosive effects tests of various materials conducted in conjunction with the developments described in [3] or related projects.

Design stresses

Table 5.3 summarizes the relevant dynamic stresses f_{dc} and f_{ds} to be used in the design of reinforced concrete elements.

Idealization of structural response

The structural response of a reinforced concrete element subjected to flexure may be represented by the idealized resistance–deflection function shown in Fig. 5.6, where r_u is the unit ultimate dynamic resistance

Plate 16. *Survivability of steel-framed buildings*

BLAST EFFECTS ON BUILDINGS

Plate 17. Severe damage to both the structure's steel frame and its lightweight metal cladding: Boucher Street, Belfast

Table 5.3. Dynamic design stresses for reinforced concrete

Type of stress	Protection category	Dynamic design stress	
		Concrete f_{dc}	Reinforcing bars f_{ds}
Bending	1	f_{dcu}	f_{dy}
	2	f_{dcu}	$f_{dy} + (f_{du} - f_{dy})/4$
Shear	1	f_{dcu}	f_{dy}
	2	f_{dcu}	f_{dy}
Compression	1 and 2	f_{dcu}	f_{dy}

determined using plastic theory, modified to account for static loads present at the time of the blast loading, X_E is the deflection at the limit of elastic behaviour, K_E is the elastic stiffness and X_m is the maximum permitted deflection corresponding to the limiting support rotation, θ, given in Table 5.2 for the appropriate protection category.

DESIGN OF ELEMENTS IN REINFORCED CONCRETE AND STRUCTURAL STEEL

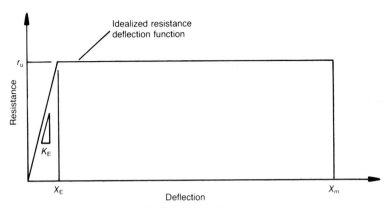

Fig. 5.6. Idealized resistance–deflection curve [3]

Response to impulse

Idealization of blast load. The blast load may be idealized to a triangular pressure–time function with zero rise time as shown in Fig. 5.7. In the impulsive regime, the duration of the applied load, t_d, is short in relation to the response time, t_m, of the element (the time for the element to attain a deflection X_m), such that $t_m/t_d \geq 3$. The loading is assumed to be uniformly distributed and is represented by the specific impulse, i.

Design for flexure.

(a) *Design objective.* To provide flexural strength and ductility so that the kinetic energy delivered by the impulsive load may be resisted by the strain energy developed by the member in deflecting to X_m.

(b) *Basic impulse equation.* For support rotations, $\theta < 5°$, the elastic part of the idealized resistance–deflection function must be considered such that

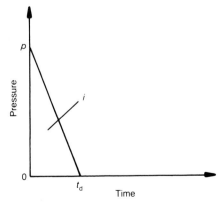

Fig. 5.7. Idealization of blast load [3]

BLAST EFFECTS ON BUILDINGS

$$\frac{i^2}{2K_{LM}m} = \frac{r_u X_E}{2} + r_u(X_m - X_E)$$

where K_{LM} is the appropriate load–mass factor obained from Table B.1 (in Appendix B) and m is the unit mass of the element.

(c) *Design steps.*

Step 1. Define the resistance–deflection function in terms of:
(i) $r_u = f(M_p, L)$ (see Table B.3), where

$$M_p = \frac{A_s f_{ds}}{b} d_c$$

for a type 2 section and

$$\rho_s = \frac{A_s}{bd_c}$$

where ρ_s is the steel ratio and hence $M_p = \rho_s f_{ds} d_c^2$.
(ii) $X_m = f(\theta)$.
(iii) $K_E = f(E, I, L)$ (see Table B.4), where $I = Fbd_c^3$ (see Fig. 5.8) and E is the modulus of elasticity of concrete.
(iv) $X_E = r_u/K_E$.

Step 2. Determine K_{LM} (see Table B.1) and $m = \rho d_c$, where ρ is the density of concrete.

Step 3. Solve for d_c based on assumed value of ρ_s.

Step 4. Calculate $t_m \simeq i/r_u$, hence t_m/t_d and check whether appropriate design procedure has been used, i.e. quasi-static/dynamic or impulse.

Design for shear. After the flexural design of the element has been completed, the required quantity of shear reinforcement must be determined. The ultimate shear is developed when the resistance reaches the ultimate value, r_u, and hence the shear reinforcement is a function of the resistance of the element and not of the applied load.

There are two critical locations where shear must be considered in the design of reinforced concrete elements. The ultimate shear stress, v_u, is calculated at a distance d_c from the supports to check the diagonal tension stress and to provide shear reinforcement in the form of stirrups, as necessary. The direct shear force or the ultimate support shear, V_s, is calculated at the face of the support to determine the required quantity of diagonal bars.

(a) *Ultimate shear stresses.* Values of ultimate shear stresses, v_u, at distance d_c from the face of the support are given in Table B.5.
(b) *Shear capacity of concrete.* The capacity of the concrete, v_c, may be taken from BS 8110: Part 1 Table 3.9 with the partial safety factor, $\gamma_m = 1 \cdot 25$, removed.

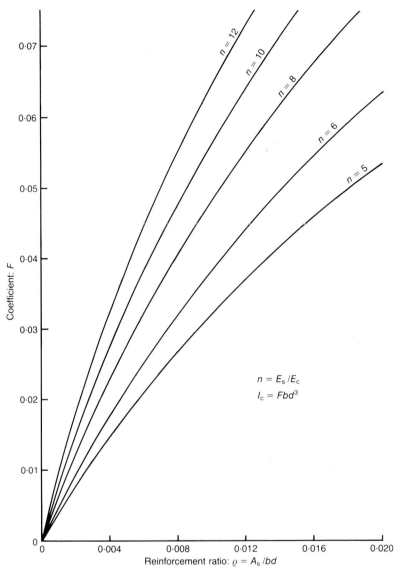

Fig. 5.8. *Coefficient for moment of inertia of cracked sections with equal reinforcement on opposite faces [3]*

(c) *Design of shear reinforcement.* Whenever the ultimate shear stress, v_u, exceeds the shear capacity of the concrete, v_c, shear reinforcement must be provided to carry the excess. The required area of stirrups, A_v, is calculated from

$$A_v = \frac{(v_u - v_c)bs}{f_{ds}}$$

where s is the spacing of stirrups in the direction parallel to the longitudinal reinforcement. For type 2 sections, the minimum design stress, $(v_u - v_c)$, to be used in the calculation of shear reinforcement is $0.85 v_c$.

(d) *Ultimate support shears.* Values of ultimate support shears, V_s, at the supports are given in Table B.6.

(e) *Direct shear capacity of concrete.* For type 2 sections where the design support rotation, θ, exceeds $2°$, the ultimate direct shear capacity of the concrete, V_d, is zero and diagonal bars are required to take all direct shear.

(f) *Design of diagonal bars.* The required area of diagonal bars at $45°$, A_d, is determined from

$$A_d = V_s b / f_{ds}$$

Response to quasi-static/dynamic loading

Idealization of blast load. The blast load may be idealized into a triangular pressure–time function with zero rise time as shown in Fig. 5.7 or to other idealizations for which response charts based on SDOF analyses are available. These may include square pulses with zero rise time, gradually applied loads, or triangular pulses with a finite rise time (see Appendix C). In the quasi-static/dynamic response regimes, the duration of the applied load, t_d, is long in relation to the response time of the element, t_m, such that $t_m / t_d < 3$. The loading is assumed to be uniformly distributed and is represented by the pressure, p, which varies with time, t.

Design for flexure.

(a) *Design objective.* To provide flexural strength and ductility such that the work done by the applied blast load may be resisted by the strain energy developed by the member in deflecting to X_m.

(b) *Design steps.*

Step 1. Define resistance–deflection function in terms of:
(i) $r_u = f(M_p, L)$ (see Table B.3),

where

$$M_p = \frac{A_s f_{ds}}{b} (d - 0.45 x) \text{ for a type 1 section and}$$

$$x = \frac{A_s f_{ds}}{0.6 b f_{dc}}$$

is the depth from the compression face to the neutral axis and d is the effective depth of the tension reinforcement and $\rho_s = (A_s / bd)$.

(ii) $X_m = f(\theta)$.

DESIGN OF ELEMENTS IN REINFORCED CONCRETE AND STRUCTURAL STEEL

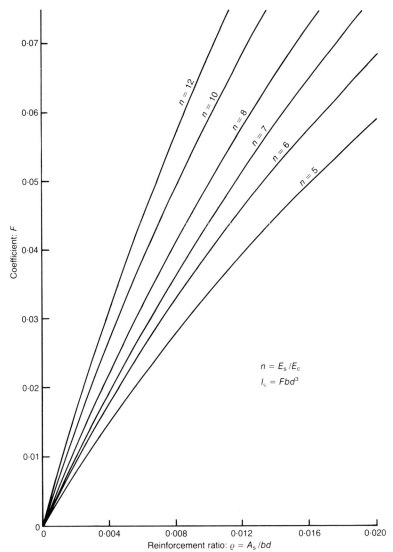

Fig. 5.9. *Coefficient for moment of inertia of cracked sections with tension reinforcement only* [3]

(iii) $K_E = f(E, I, L)$ (see Table B.4), where $I = Fbd^3$ (see Fig. 5.9).
(iv) $X_E = r_u / K_E$.

Step 2. Calculate natural period of element

$$T = 2\pi \sqrt{\frac{K_{LM} m}{K_E}}$$

where K_{LM} is the appropriate load–mass factor from Table B.1 and m is the unit mass of the element.

Step 3. Refer to appropriate SDOF response chart in Appendix C for an elasto-plastic system under idealized load to obtain:
(i) $\mu = X_m/X_E$, hence X_m and θ
(ii) t_m/t_d and hence check whether the appropriate design procedure has been used, i.e. quasi-static/dynamic or impulse.

Design for shear. The design for shear of type 1 reinforced concrete elements responding in the pressure–time regime is similar to that for type 2 elements responding to impulse with the following exceptions:

(a) *Ultimate shear stresses.*
  (i) Unless specifically required as shear reinforcement, stirrups are not required in slabs for which $\theta \leq 1°$.
  (ii) The values of ultimate shear stresses should be calculated at distance, d, from the supports, rather than d_c as given in Table B.5.
(b) *Ultimate support shears.*
  (i) For type 1 sections where $\theta \leq 2°$, the ultimate direct shear force that can be resisted by the concrete, V_d, is given by
$$V_d = 0 \cdot 15 f_{dc} bd$$
  (ii) The corresponding area of 45° diagonal bars, A_d, required becomes
$$A_d = \frac{V_s b - V_d}{f_{ds}}$$

Dynamic reactions

When a reinforced concrete element is loaded dynamically, the loads transferred to the supports are known as the dynamic reactions. The magnitude of these reactions, which may be used as the basis of design of supporting elements, is a function of both the total resistance, R, and the total load, P, applied to the element, both of which vary with time. They may be expressed generally in the form $V = \alpha R + \beta P$. Table B.7 provides values of α and β for differing support and loading arrangements.

It should be noted that the procedures described in this chapter are largely concerned with the design of elements. This is likely to be the critical design consideration when dealing with relatively large quantities of explosive at close proximity to a building structure. In situations where the overall stability of the building becomes critical, the designer will need to consider the provision of stabilizing elements such as shear walls. The external loading on complete structures and the associated blast wave–structure interaction is fully described in Chapter 3.

DESIGN OF ELEMENTS IN REINFORCED CONCRETE AND STRUCTURAL STEEL

Design example 1: reinforced concrete cantilever to resist impulsive load

A cantilever blast wall is to be designed in reinforced concrete to withstand the impulse due to the detonation of 100 kg TNT at ground level at a stand-off of 4 m.

Take the wall height to be 3 m and assume a wall symmetrically reinforced with grade 460 steel such that $\rho_s = 0 \cdot 5\%$. The concrete may be taken as grade 40 with a density of 2400 kg/m³.

Idealization of blast load. For a hemispherical surface burst, the relevant blast resultants at the wall are:

- positive phase duration, $T_s = t_d = 4 \cdot 83$ ms (Fig. 3.2)
- reflected impulse, $i_r = 5030$ kPa-ms (kN-ms/m²)⁵

Design for flexure. For a unit width of cantilever retaining wall of height, H, effective depth, d_c and a type 2 section:

(a) *Design steps.*

Step 1.

(i) $r_u = \dfrac{2M_n}{H^2}$ (see Table B.3)

where

$$M_n = \rho_s f_{ds} d_c^2$$

hence

$$r_u = \dfrac{2}{H^2} (\rho_s f_{ds} d_c^2)$$

Now $f_{ds} = f_{dy} + (f_{du} - f_{dy})/4$ (see Table 5.3)

where

$f_{dy} = 1 \cdot 20 \times 460 = 552$ N/mm² and
$f_{du} = 1 \cdot 05 \times 550 = 578$ N/mm² (see Table 5.1). Hence
$f_{ds} = 560$ N/mm² $= 560 \times 10^6$ N/m². Thus

$$r_u = \dfrac{2}{3 \cdot 0^2} (0 \cdot 005 \times 560 \times 10^6) d_c^2 \text{ N/m}^2$$

$$= 622 \times 10^3 d_c^2 \text{ N/m}^2$$

(ii) For protection category 2

$X_m = H \tan 4° = 0 \cdot 21$ m (see Table 5.2)

(iii) $K_E = 8EI/H^4$ (see Table B.4)

85

Take

$E_s = 200 \text{ kN/mm}^2 = 200 \times 10^9 \text{ N/m}^2$

$E_c = 28 \text{ kN/mm}^2 = 28 \times 10^9 \text{ N/m}^2$

$n = 200/28 = 7 \cdot 14$ and

$\rho_s = 0 \cdot 005$. Hence

$I = 0 \cdot 0245 b d_c^3$ (see Fig. 5.8). Thus

$$K_E = \left(\frac{8 \times 28 \times 10^9 \times 0 \cdot 0245 \times 1}{3 \cdot 0^4} \right) d_c^3 \text{ N/m}^2/\text{m}$$

$= 67 \cdot 8 \times 10^6 d_c^3 \text{ N/m}^2/\text{m}$

(iv) $X_E = r_u / K_E$

$$= \frac{622 \times 10^3 d_c^2}{67 \cdot 8 \times 10^6 d_c^3} = \frac{9 \cdot 2 \times 10^{-3}}{d_c} \text{ m}$$

Step 2. $K_{LM} = 0 \cdot 66$ (see Table B.1) $m = \rho d_c = 2400 d_c \text{ kg/m}^2$.

Step 3. The basic impulse equation from page 80 is

$$\frac{i^2}{2K_{LM}m} = \frac{r_u X_E}{2} + r_u(X_m - X_E)$$

$$= r_u \left(X_m - \frac{X_E}{2} \right)$$

For $i = 5030 \text{ kPa-ms} = 5030 \text{ N-s/m}^2$ we obtain

$$5030^2 = 2 \times 0 \cdot 66 \times 2400 d_c \times 622 \times 10^3 d_c^2 \left(0 \cdot 21 - \frac{9 \cdot 2 \times 10^{-3}}{2 d_c} \right)$$

and simplifying gives

$413 \cdot 8 \times 10^6 d_c^3 - 9 \cdot 06 \times 10^6 d_c^2 = 25 \cdot 3 \times 10^6$

which gives $d_c = 0 \cdot 398$ m, say 400 mm.

$A_s = 0 \cdot 005 \times 400 \times 1000 = 2000 \text{ mm}^2/\text{m}$, therefore use T20 bars at 150 mm centres on each face. Hence, the overall section thickness, using 40 mm cover, is

$T_c = 40 + 10 + 400 + 10 + 40 = 500 \text{ mm}$

Step 4. $t_m \simeq i/r_u$. Now $r_u = 622 \times 10^3 \times 0 \cdot 4^2 = 99\,520 \text{ N/m}^2$, hence

$$t_m = \frac{5030}{99\,520} = 0 \cdot 0505 \text{ s} = 50 \cdot 5 \text{ ms which gives}$$

$t_m/t_d = 50 \cdot 5/4 \cdot 83 = 10 \cdot 5 \geq 3$

Therefore impulsive loading design is valid.

Design for shear

(a) The ultimate shear stress, distance $d_c = 400$ mm from support, is given as

$$v_u = \frac{r_u(L - d_c)}{d_c}$$

$$= 99\,520(3 \cdot 0 - 0 \cdot 4)/0 \cdot 4$$

$$= 646\,900 \text{ N/m}^2 = 0 \cdot 65 \text{ N/mm}^2$$

(see Table B.5).

(b)
$$\frac{100A_s}{bd_c} = 0 \cdot 52, \; f_{cu} = 40 \text{ N/mm}^2$$

Therefore

$$v_c = 0 \cdot 58 \text{ N/mm}^2 \quad (\text{BS 8110, Part I: Table 3.9}),$$

$$(v_u - v_c) = 0 \cdot 07 \text{ N/mm}^2 \times (0 \cdot 85 v_c = 0 \cdot 49 \text{ N/mm}^2).$$

(c) The required area of stirrups of width, $b = 150$ mm and spacing, $s = 300$ mm

$$A_v = \frac{(v_u - v_c)bs}{f_{ds}}$$

Using grade 250 steel, $f_{ds} = f_{dy} = 1 \cdot 1 \times 250 = 275 \text{ N/mm}^2$ (see Tables 5.1 and 5.3), hence

$$A_v = \frac{0 \cdot 49 \times 150 \times 300}{275} = 80 \cdot 2 \text{ mm}^2$$

Therefore use R10 stirrups at 300 mm centres.

(d) The ultimate support shear is given by

$$V_s = r_u L$$
$$= 99\,520 \times 3 \cdot 0$$
$$= 298\,600 \text{ N/m} = 299 \text{ N/mm} \quad (\text{see Table B.6})$$

(e) $V_d = 0$

(f) The required area of diagonal bars at 45° and spacing of $b = 150$ mm; $A_d = (V_s b/f_{ds})$.

Using grade 460 steel, $f_{ds} = f_{dy} = 1 \cdot 1 \times 460 = 506 \text{ N/mm}^2$ (Tables 5.1 and 5.3), hence

$$A_d = 299 \times 150/506 = 88 \cdot 2 \text{ mm}^2$$

Therefore use one row of T12 bars at 150 mm centres or two rows of T8 bars at 150 mm centres.

The designer should now proceed to check the overall stability of the wall and the detailed design of the foundations.

Design example 2: reinforced concrete wall panel to resist quasi-static/dynamic load

The flexural capacity of a fixed ended reinforced concrete wall panel, 600 mm thick and spanning vertically over an effective height of 7 m, is to be checked for protection category 1 against a specified blast loading threat. The wall is reinforced symmetrically with grade 460 bars such that $\rho_s = \rho'_s = 0.35\%$, $d = 550$ mm and $d' = 50$ mm. The concrete is grade 40.

Idealization of blast load. The specified blast loading threat may be idealized to a triangular pressure–time function with $p_r = 220$ kPa and $t_d = 60$ ms.

Design for flexure. For a unit width of wall of effective height, H, effective depth, d, and a type 1 section.

(a) *Design steps.*

Step 1.

(i) $r_u = \dfrac{8(M_n + M_p)}{H^2}$ (see Table B.3), where

$$M_n = M_p = \frac{A_s f_{ds}}{b}(d - 0.45x)$$

and

$$x = \frac{A_s f_{ds}}{0.6 b f_{dc}}$$

for a symmetrically reinforced type 1 section in which the contribution of the 'compression reinforcement' is ignored. Now

$f_{ds} = f_{dy} = 1.2 \times 460 = 552$ N/mm² (Tables 5.1 and 5.3)

and $f_{dc} = f_{dcu} = 1.25 \times 40 = 50$ N/mm² (Tables 5.1 and 5.3)

Also $A_s = A'_s = 0.35 \times 1000 \times 550/100$ mm²/m
$= 1925$ mm²/m

Therefore, $x = 1925 \times 552/(0.6 \times 1000 \times 50$ mm$)$
$= 35.5$ mm

(this justifies ignoring the contribution of the 'compression reinforcement') and

$M_n = M_p = 1925 \times 552 (550 - 0.45 \times 35.5) \times 10^{-3}$ Nm/m
$= 566\,000$ Nm/m

Hence $r_u = 8 \times 2 \times 566\,000/7.0^2 = 184\,800$ N/m²

(ii) For protection category 1, θ is limited to 2° (Table 5.2). In the absence of stirrups, θ is further limited to 1°.

Therefore, assume that $X_m = 3.5 \tan 1° = 0.0611$ m $= 61.1$ mm

(iii) $K_E = 307EI/H^4$ (see Table B.4)

As in Example 1, $n = 7\cdot14$. Also, $\rho_s = 0\cdot0035$

Hence $I = 0\cdot018bd^3$ (Fig. 5.9).

Thus

$$K_E = \frac{307 \times 28 \times 10^9 \times 0\cdot018 \times 1 \times 0\cdot55^3}{7\cdot0^4}$$

$$= 10\cdot72 \times 10^6 \, \text{N/m}^2/\text{m}$$

(iv) $X_E = r_u/K_E$
$= 184\,800/10\cdot72 \times 10^6 = 0\cdot0172 \, \text{m}$
$= 17\cdot2 \, \text{mm}$

Step 2.

$$T = 2\pi \sqrt{\frac{K_{LM}m}{K_E}}$$

Now

$$K_{LM} \approx \frac{0\cdot77 + 0\cdot66}{2} = 0\cdot72$$

and $m = 2400 \times 0\cdot6 = 1440 \, \text{kg/m}^2$. Therefore

$$T = 2\pi \left(\frac{0\cdot72 \times 1440}{10\cdot72 \times 10^6}\right)^{1/2} = 0\cdot0618 \, \text{s}$$

$$= 61\cdot8 \, \text{ms}$$

Step 3. Referring to the SDOF response chart C.1, $r_u/P = 184\,800/220\,000 = 0\cdot84$ and $t_d/T = 60/61\cdot8 = 0\cdot97$. Hence, $X_m/X_E = 2\cdot7$ and $X_m = 2\cdot7 \times 17\cdot2 = 46\cdot4 \, \text{mm} < 61\cdot1 \, \text{mm}$ which is satisfactory. Referring to the SDOF response chart C.2, $t_m/t_d = 0\cdot67 < 3$. Therefore quasi-static/dynamic loading design is valid and the flexural capacity is adequate. Further checks may be necessary for shear capacity and the magnitude of dynamic reactions applied to the top and bottom supporting elements.

Introduction to the design of structural steel elements to resist blast loading

Design stresses

Table 5.4 summarizes the relevant dynamic stresses, f_{ds} and f_{dv} for flexure and shear, respectively, to be used in the design of structural steel elements.

Table 5.4. Dynamic design stresses for structural steel

Type of stress	Protection category	Dynamic design stress
Bending	1	$f_{ds} = f_{dy}$
	2	$f_{ds} = f_{dy} + (f_{du} - f_{dy})/4$
Shear	1 and 2	$f_{dv} = 0 \cdot 55 f_{ds}$

Idealization of structural response

The structural response of a steel element subjected to flexure may be represented by the idealized resistance–deflection function shown in Fig. 5.6, where r_u is the unit ultimate dynamic resistance determined using plastic theory modified to account for static loads present at the time of the blast loading; X_E is the deflection at the limit of elastic behaviour; K_E is the elastic stiffness and X_m is the maximum permitted deflection corresponding to the more critical of the limiting support rotation, θ, or ductility ratio, μ, given in Table 5.2 for the appropriate protection category.

Response to quasi-static/dynamic loading

Idealization of blast load. The blast load may be idealized into a triangular pressure–time function with zero rise time as illustrated in Fig. 5.7 or to other idealizations for which response charts based upon SDOF analyses are available (see Appendix C).

Design for flexure.

(a) *Design objective.* To provide flexural strength and ductility such that the work done by the applied blast load may be resisted by the strain energy developed by the member in deflecting to X_m.

(b) *Design steps.*

   Step 1. Carry out preliminary design assuming an equivalent static ultimate resistance as defined in Table 5.5.

   (i) Determine the required resistance, r_u, using Table 5.5.
   (ii) Determine the required plastic moment of resistance, $M_p = f(r_u, L)$ (use Table B.3).

Table 5.5. Equivalent static ultimate resistance for preliminary design of steel elements in flexure

Protection category	Equivalent static ultimate resistance
1	$r_u = 1 \cdot 0 \, p_{max}$
2	$r_u = 0 \cdot 5 \, p_{max}$

(iii) Select a steel member using appropriate relationship between M_p, f_{ds}, S and Z.

Step 2. Calculate the natural period of the element using the following equation

$$T = 2\pi \sqrt{\frac{K_{LM} m}{K_E}}$$

where K_{LM} is the appropriate load−mass factor from Table B.1 and m is the unit mass of the element.

Step 3. Refer to appropriate SDOF response chart in Appendix C for an elasto-plastic system under idealized load to obtain

(i) $\mu = X_m/X_E$, hence X_m and θ, and
(ii) t_m/t_d and hence check whether the appropriate design procedure has been used, i.e. quasi-static/dynamic.

Check for shear and secondary effects.

(a) *Ultimate support shear.* Values of ultimate support shears, V_s, are given in Table B.6.
(b) *Ultimate shear capacity.* The ultimate shear capacity is given by $V_p = f_{dv} A_w$, where A_w is the area of the web.
(c) *Local buckling and web stiffeners.* In order to ensure that a steel beam will attain fully plastic behaviour and hence the required ductility at plastic hinge locations, it is necessary that the elements of a beam section meet the normal minimum thickness requirements sufficient to prevent a local buckling failure. Similarly, web stiffeners should be employed at locations of concentrated loads and reactions to provide a gradual transfer of forces to the web.
(d) *Lateral bracing.* Members subjected to bending about their strong axes may be susceptible to lateral-torsional buckling in the direction of their weak axes if their compression flange is not laterally braced. In order for a plastically designed member to reach its collapse mechanism, lateral supports must be provided at and adjacent to the location of plastic hinges. In designing such bracing due consideration should be given to the possibility of rebound which induces stress reversal.

Design example 3: structural steel beam to resist quasi-static/dynamic load

A simply supported steel floor beam is to be designed for protection category 1 against a specified blast-loading threat. The floor beam is one of a series of beams spaced at 1·4 m which are to support a 4 mm thick steel deck plate over an effective span of 5·2 m. Take $f_y = 275 \text{ N/mm}^2$, $E_s = 210 \text{ kN/mm}^2$ and $\rho = 7850 \text{ kg/m}^3$.

BLAST EFFECTS ON BUILDINGS

Idealization of blast load. The specified blast loading threat may be idealized to a triangular pressure−time function with $p_r = 50$ kPa and $t_d = 40$ ms.

Design for flexure.

Step 1. For protection category 1, $\theta \leq 2°$ and $\mu \leq 10$.

(i) For a preliminary design take $r_u = 1 \cdot 0\, p_r$ (see Table 5.5). Thus, the required resistance per unit length is given as

$$r_u = 1 \cdot 0 \times 50 \times 1 \cdot 4 = 70 \text{ kN/m}.$$

(ii) The required plastic moment of resistance is

$$M_p = \frac{r_u L^2}{8} = \frac{70 \times 5 \cdot 2^2}{8} \quad \text{(see Table B.3)}$$

$$= 236 \cdot 6 \text{ kN m}$$

(iii) Take $M_p = f_{ds} S$ where

$$f_{ds} = f_{dy} = 1 \cdot 1 \times 1 \cdot 20 \times 275 = 363 \text{ N/mm}^2 \text{ and}$$

$$S \geq 236 \cdot 6 \times 10^6 / 363 = 0 \cdot 652 \times 10^6 \text{ mm}^3$$

Select a 356 × 127 × 39 kg/m universal beam ($S = 0 \cdot 654 \times 10^6$ mm³, $M_p = 0 \cdot 654 \times 10^6 \times 363 = 237 \times 10^6$ N mm = 237 kN m).

Step 2.

$$K_{LM} = \frac{0 \cdot 78 + 0 \cdot 66}{2} = 0.72 \quad \text{(see Table B.1)}$$

Mass, m, per unit length $= 39 + (7850 \times 1 \cdot 4 \times 0 \cdot 004) = 83 \cdot 0$ kg/m

$$K_E = \frac{384 EI}{5 L^4} \quad \text{(see Table B.4)},$$

where $I = 100 \cdot 87 \times 10^6$ mm⁴.

Thus

$$K_E = \frac{384 \times 210 \times 10^3 \times 100 \cdot 87 \times 10^6}{5 \times (5 \cdot 2 \times 10^3)^4}$$

$$= 2 \cdot 225 \text{ N/mm/mm}$$
$$= 2 \cdot 225 \times 10^6 \text{ N/m/m}.$$

Hence

$$T = 2\pi \sqrt{\frac{K_{LM} m}{K_E}}$$

$$= 2\pi \left(\frac{0\cdot 72 \times 83\cdot 0}{2\cdot 225 \times 10^6} \right)^{1/2}$$
$$= 0\cdot 0326 \text{ s}$$
$$= 32\cdot 6 \text{ ms}$$

Step 3.

(i) Referring to the SDOF response chart C.1

$$r_u = \frac{8M_p}{L^2} = \frac{8 \times 237 \times 10^3}{5\cdot 2^2}$$
$$= 70\cdot 1 \times 10^3 \text{ N/m}$$
$$P = 50 \times 10^3 \times 1\cdot 4 = 70 \times 10^3 \text{ N/m}$$

Thus, $r_u/P = 1\cdot 00$ and $t_d/T = 40/32\cdot 6 = 1\cdot 23$. Hence, $X_m/X_E = \mu = 2\cdot 1 \le 10$, which is satisfactory.

$$X_m = \mu X_E = \mu \frac{r_u}{K_E}$$
$$= 2\cdot 1 \times 70\cdot 1 \times 10^3/2\cdot 225 \times 10^6$$
$$= 0\cdot 0662 \text{ m}$$
$$= 66\cdot 2 \text{ mm}$$
$$\tan \theta = \frac{X_m}{L/2} = 66\cdot 2 \times 2/5\cdot 2 \times 10^3$$

Thus, $\theta = 1\cdot 46° \le 2°$ which is satisfactory. The beam is slightly over-designed and there may be some scope for further refinement of the section.

(ii) Referring to SDOF response chart C.2, $t_m/t_d = 0\cdot 51 < 3$. Therefore pressure–time loading design is valid and the flexural capacity is adequate.

Check for shear and secondary effects.

(a) The ultimate support shear is

$$V_s = \frac{r_u L}{2} = 70\cdot 1 \times 10^3 \times 5\cdot 2/2$$
$$= 182\cdot 3 \times 10^3 \text{ N}$$

(b) The ultimate shear capacity is

$$f_{dv} = 0\cdot 55 f_{ds} = 0\cdot 55 \times 363$$
$$= 200 \text{ N/mm}^2$$
$$V_p = f_{dv} A_w = 200 \times 352\cdot 8 \times 6\cdot 5$$
$$= 458\cdot 6 \times 10^3 \text{ N}$$

$V_p > V_s$ which is satisfactory.

(c) Check for local buckling and provide web stiffeners at support.
(d) Top flange of beam is laterally restrained under normal downward loading. Provide lateral bracing to lower flange in the event of load reversals.

References

1. British Standards Institution. *Structural use of steelwork in building. Code of practice for design in simple and continuous construction.* BSI, London, 1985, BS 5950: Part 1.
2. British Standards Institution. *Structural use of concrete. Code of practice for design and construction.* BSI, London, 1985, BS 8110: Part 1.
3. US Department of the Army Technical Manual, TM5-1300. *Design of structures to resist the effects of accidental explosions.* 1990.
4. British Standards Institution. *Bending dimensions and scheduling of reinforcement or concrete.* BSI, London, 1981, BS 4466.
5. Hyde D.W. CONWEP: weapons effects calculation program based on US Department of the Army Technical Manual, TM5-855-1. *Fundamentals of protective design for conventional weapons.* 1992.

6 Implications for building operation

Health and safety regulations

The Management of Health and Safety at Work Regulations, 1992, which came into force on 1 January 1993, oblige an employer to establish and, where necessary, give effect to appropriate procedures to be followed in the event of serious and imminent danger to persons at work in his or her undertaking. Current legal opinion implies that this legislation may be interpreted to include the safety of employees when there is a threat of a terrorist act or other violent act perpetrated by, for example, a disgruntled employee. Further, The Construction Products Directive of the European Community contains, within Annex 1, essential requirements to ensure that products are fit for their intended use. This includes safety in use and in particular it specifies that 'the construction work must be designed and built in such a way that it does not present unacceptable risks of accidents in service or in operation such as slipping, falling, collision, burns, electrocution, *injury from explosion* [author's emphasis].' The attention of the reader is also drawn to the Construction (Design and Management) Regulations, 1994. These clarify the duties of those involved with the management of risks arising from buildings and structures throughout their life cycle. Therefore, it is imperative to have well-established and practised contingency procedures in place.

Threat assessment

The following factors should be taken into consideration in the preparation of a threat assessment for a particular building:

- Advice from the local police force which may be based upon the Home Office document *Bombs — protecting people and property* [1].
- The current national and international situation as determined from news coverage.
- Any special circumstances of the occupants.
- The building location, for example whether there is a 'high risk'

building close by, an attack on which may result in collateral damage to the building in question.

Pre-event contingency planning

When preparing the pre-event plan certain important aspects need to be considered. They fall into two categories: initiating and managing the plan in response to a threat; and the response actions relating to the various threats.

Initiating and managing the plan

It is important to appoint a single person (a security coordinator) to have full responsibility for initiating and managing the plan. He or she must receive total support from senior management. Further, all staff, including the senior management, must accept and act on this person's instructions during the emergency. The security coordinator will need deputies.

The plan should be discussed with the police and fire brigade as early as possible. The Home Office booklet *Bombs — Protecting People and Property* contains a great deal of valuable advice and is an excellent aid to drawing up contingency plans.

To ensure that the plan works satisfactorily under a real threat, and that staff appreciate what is expected of them, it must be regularly practised and further, it must be kept up to date.

Response actions

(a) *The telephoned bomb threat.* On receipt of a bomb threat call, the security coordinator should be informed of the contents of the call in order that they can make a considered assessment of the threat. Based on the result of this assessment the security coordinator will inform the police and initiate one of the following:

 (i) Do nothing.
 (ii) Undertake a limited search.
 (iii) Undertake a full search then either evacuate staff or assemble staff in the bomb shelter area (BSA).

(b) *The delivered item (letter bomb or package bomb).* If a letter or package is suspected of containing an explosive device, then the following actions should be undertaken:

 (i) Leave it alone.
 (ii) Clear the room plus the rooms either side, above and below.
 (iii) Inform the police.

Ensure that the person who discovered the suspect package remains on hand to assist the police when they arrive.

It is important that all personnel who are involved in receiving mail and delivered items should be made aware of the above procedures.

(c) *The internal improvised explosive device (IED).* If it is believed that an IED has been deposited inside the building then the following procedures for searching the building should be contained in the plan:

(i) Appointed teams to search the common public areas.
(ii) Individual staff to search their own work place.

If the suspected IED is found the area should be cleared and its location reported to the security coordinator. The security coordinator will then have to consider continuing the search of other areas since more than one device may have been planted. Also, this will assist when the decision to re-occupy the building has to be taken.

Depending upon the location of the device, consideration should be given to using the BSA or evacuating the building.

(d) *The external IED.* Typically the external IED falls into two categories: the small carried device, e.g. briefcase, carrier bag, duffle bag, etc., and the vehicle-borne device ranging from a car to a large lorry.

Generally the reaction to both the small and large device will be similar. However, when the location of the small device is known and it is assessed that there are no other devices in the area then, depending on the size and spread of the building, the reaction plan can be applied to a limited area of the building. If evacuation is the adopted policy, as the building is such that it does not offer a BSA, then the assembly points will have to be a minimum of 400 m from the building. Consideration will also have to be given to the potential deployment, by the terrorist, of a secondary device designed to cause injuries to the authorities attending the aftermath of the original device.

Further considerations

To enable the plan to be initiated and managed efficiently and effectively, communications between the security coordinator and all the staff and between the security coordinator and the police will be necessary throughout the whole of the incident. Therefore consideration should be given to the provision of a voice alarm system (VAS) within the building and a number of portable telephones for external communication.

If the BSA philosophy is adopted it is important to ensure that the control point from where the security coordinator is able to communicate with staff and police offers the same level of protection as that provided by the BSA.

To aid the retrieval of data and thus minimize the disruption to the business, consideration should be given to implementing the following at the end of each working day and whenever assembly in the BSAs or evacuation is actioned.

- Cover IT equipment, particularly the data storage systems;
- Impose a clear desk policy.

Post-event contingency planning

A post-event contingency plan is essential for business survival. It is not only necessary for dealing with terrorist outrages but also for many other forms of disaster from flooding to fire and from a major communications failure to a plane crash.

There are many commercially available pre-designed plans or consultants capable of providing a business specific plan. To assist in ensuring that the finally adopted plan is what is required, a list of critical points which should be included are set out below.

(a) A pre-identified team to manage the disaster whose initial tasks will include assessing the extent of the damage, setting an objective, e.g. 'Business as usual on Monday' and communicating with staff. The recovery procedures should be put in hand by empowerment, i.e. giving full authority to a staff member to take all reasonable measures to fulfil his or her allotted task (e.g. no requirement for three competitive quotations for the purchase of equipment, etc.), communicating with the media (proactive if possible), communicating with staff in non-affected offices, shareholders, clients, etc.
(b) A pre-identified team to continue the core business.
(c) A communication cascade system for all staff.
(d) Recovery procedures shall include the establishment of a communications centre, including a help desk and the establishment of a computer facility for data retrieval and usage.

One vitally important point is to ensure that copies of the plan are easily available in the event of a disaster. It is not sufficient to have one copy supposedly held safely in the building. Further, the plan and data must be kept up to date. Telephone numbers and names and location of staff frequently change.

Reference

1. Home Office. *Bombs — protecting people and property*. A handbook for managers and security officers, March, 1994.

Appendix A. Simplified design procedure for determining the appropriate level of glazing protection

When using this design guidance it is important to note the following points when considering the larger devices, i.e. van bombs and above.

- The structure is capable of withstanding the blast.
- At small stand-off distances the cladding, local to the device, will not survive.
- The glazing and frame is not stronger than the cladding to which it is fixed.

The simplified design procedure for determining the appropriate level of glazing protection is as follows.

(a) *Determine the threat.* Select the size of device based on the threat assessment (see Chapter 6).
(b) *Establish the stand-off distance.* Measure the stand-off distance from the perimeter of the building to the location where the selected device could be sited.
(c) *Extract the stand-off distance for internal flying glass (4 mm annealed).* From Table A.1, for the selected device size extract the stand-off that will produce internal flying glass (4 mm annealed) and hence cause personal injury.
(d) *Calculate the 'comparative value'.* The 'comparative value' is calculated by dividing the stand-off for internal glass (item c) by the measured stand-off distance (item b),

$$\text{comparative value} = \frac{\text{stand-off 4 mm annealed}}{\text{stand-off measured}}$$

(e) *Selecting the glazing.* Select appropriate glazing by matching the calculated 'comparative value' (item d) with the values given in Table A.2. Note that if laminated glass, single or double, is selected then the use of the frame and fixing design parameters in Table A.3 is necessary.

BLAST EFFECTS ON BUILDINGS

Table A.1. Stand-off distances to produce internal flying glass

Device	Stand-off for 4 mm annealed glass: m
Small package	10
Small briefcase	14
Large briefcase	20
Suitcase	26
Car	60
Small van*	120
Large van	140
Small lorry†	160
Large lorry	200

Note. Stand-off distance is the distance from the centre of the explosion to the object under consideration (in this case the glazing). These distances are approximate. In certain circumstances, such as where the configuration of surrounding buildings causes a local concentration of the blast effect, the distances could be increased by up to 30%.
* St Mary Axe, City of London device.
† Bishopsgate, City of London device.

Table A.2. Internal flying glass. Comparison of levels of protection provided by various glass types and applied protective measures

Glazing type	Comparative value	Glazing type	Comparative value
Annealed glass (A)		*Laminated glass (L)‡*	
4 mm	1·0	6·4 mm or 6·8 mm	2·5
4 mm + ASF*	1·7	7·5 mm	2·9
4 mm + ASF + BBNC†	2·0	11·5 mm	3·3
Toughened glass (T)		*Double glazed units‡*	
6 mm	2·0	6 mm T + 6 mm T	2·5
6 mm + ASF	2·5	6 mm A + 7·5 mm L	3·3
8 mm	2·5	6 mm T + 7·5 mm L	4·0
8 mm + ASF	2·9		
10 mm	2·9		
10 mm + ASF	3·3		

i.e. 6·4 mm laminated glass is 2·5 times better than 4 mm annealed glass.

* ASF — anti-shatter film, 100 μm thick.
† BBNC — bomb blast net curtains.
‡ For laminated glass to be effective it must be securely fixed into its frames preferably with structural silicon and the frames must have glazing rebates of a minimum of 25 mm deep. However, in special frames with deeper rebates (35 mm) the percentage improvement will increase. The frames in turn must be securely fixed to the surrounding structure.

Table A.3. Frame and fixing design parameters for laminated glass

Laminated glass thickness: mm	Approx. glass pane size: m²	Equivalent ultimate static load: kN/m²	Minimum glazing rebate: mm
6·4	0·6	6·0	25
	1·8	3·0	25
	3·0	3·0	30
6·8	0·6	8·0	25
	1·8	4·0	25
	3·0	4·0	30
7·5	0·6	12·0	25
	1·8	7·0	30
	3·0	6·0	30
11·5	0·6	18·0	25
	1·8	11·0	30
	3·0	9·0	30

The frames and fixings should be designed to accommodate two modes of action occurring simultaneously:
(i) A line load of F kN/m acting on the frame perpendicular to the plane of the window.
(ii) A line load of $0·5F$ kN/m acting on the frame in the plane of the window, towards the centre of the pane.
F is defined as the average edge reaction, i.e. $F = (WA/P)$, where W = the equivalent ultimate static load, kN/m², A = the area of the window, m², and P = the perimeter of the window, m.
The above figures for equivalent static loads are a guide only.

(f) *Worked example.*

    (i) Threat assessment determines a small van-sized device.
    (ii) The closest a vehicle can get to the building is measured as 30 m.
    (iii) The stand-off for 4 mm annealed for a small van is 120 m.
    (iv) The calculated comparative value is 120/30 = 4·0.
    (v) The glazing type with a 4·0 or more comparative value is:

        6 mm toughened + 7·5 mm laminated.

Note that frames and fixings are to be designed using the appropriate loading parameters.

Appendix B. Transformation factors for beams and one-way slabs*

---

* Refer to Chapter 5 for notation.

APPENDIX B

Table B.1. Transformation factors for one-way elements (after US Department of the Army Technical Manual, TM5-1300. Design of structures to resist the effects of accidental explosions, 1990)

Edge conditions and loading diagrams	Range of behaviour	Load factor K_L	Mass factor K_M	Load–mass factor K_{LM}
Simply supported, uniform load, span L	Elastic	0·64	0·50	0·78
	Plastic	0·50	0·33	0·66
Simply supported, point load P at mid-span ($L/2$, $L/2$)	Elastic	1·0	0·49	0·49
	Plastic	1·0	0·33	0·33
Fixed–simple, uniform load, span L	Elastic	0·58	0·45	0·78
	Elasto-plastic	0·64	0·50	0·78
	Plastic	0·50	0·33	0·66
Fixed–simple, point load P at mid-span	Elastic	1·0	0·43	0·43
	Elasto-plastic	1·0	0·49	0·49
	Plastic	1·0	0·33	0·33
Fixed–fixed, uniform load, span L	Elastic	0·53	0·41	0·77
	Elasto-plastic	0·64	0·50	0·78
	Plastic	0·50	0·33	0·66
Fixed–fixed, point load P at mid-span	Elastic	1·0	0·37	0·37
	Plastic	1·0	0·33	0·33
Cantilever, uniform load, span L	Elastic	0·40	0·26	0·65
	Plastic	0·50	0·33	0·66
Cantilever, point load P at tip	Elastic	1·0	0·24	0·24
	Plastic	1·0	0·33	0·33
Simply supported, two point loads $P/2$ at third points ($L/3$, $L/3$, $L/3$)	Elastic	0·87	0·52	0·60
	Plastic	1·0	0·56	0·56

Table B.2. Elastic and elasto-plastic unit resistances for one-way elements (as per Table B.1)

Edge conditions and loading diagrams	Elastic resistance, r_e	Elasto-plastic resistance, r_{ep}
simply supported, UDL, span L	r_u	—
simply supported, point load P at mid-span, $L/2 + L/2$	R_u	—
fixed-pinned, UDL, span L	$\dfrac{8M_N}{L^2}$	r_u
fixed-pinned, point load P at mid-span, $L/2 + L/2$	$\dfrac{16M_N}{3L}$	R_u
fixed-fixed, UDL, span L	$\dfrac{12M_N}{L^2}$	r_u
fixed-fixed, point load P at mid-span, $L/2 + L/2$	$\dfrac{8M_N}{L}$	R_u
cantilever, UDL, span L	r_u	—
cantilever, point load P at tip, span L	R_u	—
simply supported, two point loads $P/2$ at third points, $L/3 + L/3 + L/3$	R_u	—

APPENDIX B

Table B.3. Ultimate unit resistances for one-way elements (as per Table B.1)

Edge conditions and loading diagrams	Ultimate resistance
Simply supported, uniform load, span L	$r_u = \dfrac{8M_p}{L^2}$
Simply supported, point load P at $L/2$	$R_u = \dfrac{4M_p}{L}$
Fixed-simple, uniform load, span L	$r_u = \dfrac{4(M_N + 2M_p)}{L^2}$
Fixed-simple, point load P at $L/2$	$R_u = \dfrac{2(M_N + 2M_p)}{L}$
Fixed-fixed, uniform load, span L	$r_u = \dfrac{8(M_N + M_p)}{L^2}$
Fixed-fixed, point load P at $L/2$	$R_u = \dfrac{4(M_N + M_p)}{L}$
Cantilever, uniform load, span L	$r_u = \dfrac{2M_N}{L^2}$
Cantilever, point load P at tip	$R_u = \dfrac{M_N}{L}$
Simply supported, two point loads $P/2$ at $L/3$	$R_u = \dfrac{6M_p}{L}$

Table B.4. Elastic, elasto-plastic and equivalent elastic stiffnesses for one-way elements (as per Table B.1)

Edge conditions and loading diagrams	Elastic stiffness, K_e	Elasto-plastic stiffness, K_{ep}	Equivalent elastic stiffness K_E
Simply supported, uniform load, span L	$\dfrac{384EI}{5L^4}$	—	$\dfrac{384EI}{5L^4}$
Simply supported, point load P at midspan ($L/2$, $L/2$)	$\dfrac{48EI}{L^3}$	—	$\dfrac{48EI}{L^3}$
Fixed-simply supported, uniform load, span L	$\dfrac{185EI}{L^4}$	$\dfrac{384EI}{5L^4}$	$\dfrac{160EI^*}{L^4}$
Fixed-simply supported, point load P at midspan	$\dfrac{107EI}{L^3}$	$\dfrac{48EI}{L^3}$	$\dfrac{106EI^*}{L^3}$
Fixed-fixed, uniform load, span L	$\dfrac{384EI}{L^4}$	$\dfrac{384EI}{5L^4}$	$\dfrac{307EI^*}{L^4}$
Fixed-fixed, point load P at midspan	$\dfrac{192EI}{L^3}$	$\dfrac{48EI\dagger}{L^3}$	$\dfrac{192EI^*}{L^3}$
Cantilever, uniform load, span L	$\dfrac{8EI}{L^4}$	—	$\dfrac{8EI}{L^4}$
Cantilever, point load P at tip	$\dfrac{3EI}{L^3}$	—	$\dfrac{3EI}{L^3}$
Simply supported, two point loads $P/2$ at third points	$\dfrac{56 \cdot 4EI}{L^3}$	—	$\dfrac{56 \cdot 4EI}{L^3}$

* Valid only if $M_N = M_p$.
† Valid only if $M_N < M_p$.

Table B.5. Ultimate shear stress at distance d_c from face of support for one-way elements (as per Table B.1)

Edge conditions and loading diagrams	Ultimate shear stress, v_u
Simply supported, uniform load over L	$\dfrac{r_u(\tfrac{1}{2}L - d_c)}{d_c}$
Simply supported, point load P at midspan ($L/2$, $L/2$)	$\dfrac{R_u}{2d_c}$
Fixed-simple, uniform load over L	Left support $\; r_u\left(\dfrac{5L}{8} - d_c\right)/d_c$ Right support $\; r_u\left(\dfrac{3L}{8} - d_c\right)/d_c$
Fixed-simple, point load P at midspan	Left support $\; \dfrac{11 R_u}{16 d_c}$ Right support $\; \dfrac{5 R_u}{16 d_c}$
Fixed-fixed, uniform load over L	$\dfrac{r_u(\tfrac{1}{2}L - d_c)}{d_c}$
Fixed-fixed, point load P at midspan	$\dfrac{R_u}{2d_c}$
Cantilever, uniform load over L	$\dfrac{r_u(L - d_c)}{d_c}$
Cantilever, point load P at tip	$\dfrac{R_u}{d_c}$
Simply supported, two point loads $P/2$ at third points ($L/3$, $L/3$, $L/3$)	$\dfrac{R_u}{2d_c}$

Table B.6. Support shears for one-way elements (as per Table B.1)

Edge conditions and loading diagrams	Support reactions, V_s
Simply supported, uniform load, span L	$\dfrac{r_u L}{2}$
Simply supported, point load P at $L/2$	$\dfrac{R_u}{2}$
Fixed-simple, uniform load, span L	Left reaction $\dfrac{5 r_u L}{8}$ Right reaction $\dfrac{3 r_u L}{8}$
Fixed-simple, point load P at $L/2$	Left reaction $\dfrac{11 R_u}{16}$ Right reaction $\dfrac{5 R_u}{16}$
Fixed-fixed, uniform load, span L	$\dfrac{r_u L}{2}$
Fixed-fixed, point load P at $L/2$	$\dfrac{R_u}{2}$
Cantilever, uniform load, span L	$r_u L$
Cantilever, point load P at tip	R_u
Simply supported, two point loads $P/2$ at $L/3$	$\dfrac{R_u}{2}$

108

APPENDIX B

Table B.7. Dynamic reactions for one-way elements (after Biggs J.M. Introduction to structural dynamics. McGraw-Hill, New York, 1964)

Edge conditions and loading diagrams	Dynamic reactions V		
	Elastic	Elasto-plastic	Plastic
Simply supported, uniform load $P = pL$	$0\cdot39R + 0\cdot11P$	—	$0\cdot38R + 0\cdot12P$
Simply supported, point load at $L/2$	$0\cdot78R - 0\cdot28P$	—	$0\cdot75R - 0\cdot25P$
Fixed-simple, uniform load $P = pL$	Left reaction $0\cdot43R + 0\cdot19P$ Right reaction $0\cdot26R + 0\cdot12P$	$0\cdot39R + 0\cdot11P$ $\pm M_N/L$	$0\cdot38R + 0\cdot12P$ $\pm M_N/L$
Fixed-simple, point load at $L/2$	Left reaction $0\cdot54R + 0\cdot14P$ Right reaction $0\cdot25R + 0\cdot07P$	$0\cdot78R - 0\cdot28P$ $\pm M_N/L$	$0\cdot75R - 0\cdot25P$ $\pm M_N/L$
Fixed-fixed, uniform load $P = pL$	$0\cdot36R + 0\cdot14P$	$0\cdot39R + 0\cdot11P$	$0\cdot38R + 0\cdot12P$
Fixed-fixed, point load at $L/2$	$0\cdot71R - 0\cdot21P$	—	$0\cdot75R - 0\cdot25P$
Cantilever, uniform load $P = pL$			
Cantilever, point load at L			
Simply supported, two point loads $P/2$ at $L/3$	$0\cdot53R - 0\cdot03P$	—	$0\cdot52R - 0\cdot02P$

Appendix C. Maximum deflection and response time for elasto-plastic single degree of freedom systems*

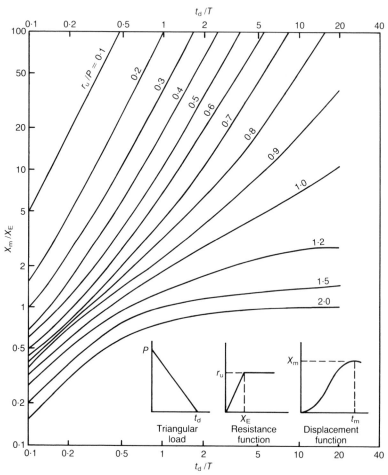

Fig. C.1. Maximum deflection of elasto-plastic, one degree of freedom system for triangular load (after US Department of the Army Technical Manual, TM5-1300. Design of structures to resist the effects of accidental explosions, 1990)

* Refer to Chapter 5 for notation.

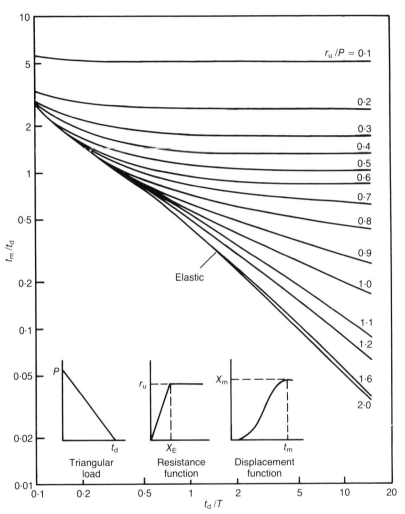

Fig. C.2. Maximum response time of elasto-plastic, one degree of freedom system for triangular load (as per Fig. C.1.)

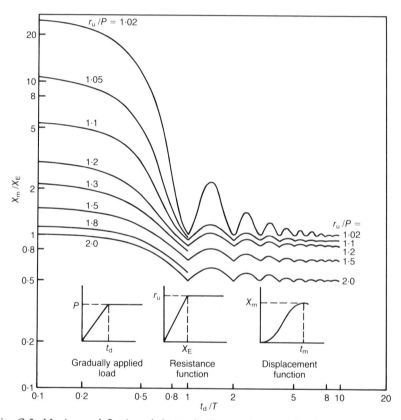

Fig. C.3. Maximum deflection of elasto-plastic, one degree of freedom system for gradually applied load (as per Fig. C.1.)

APPENDIX C

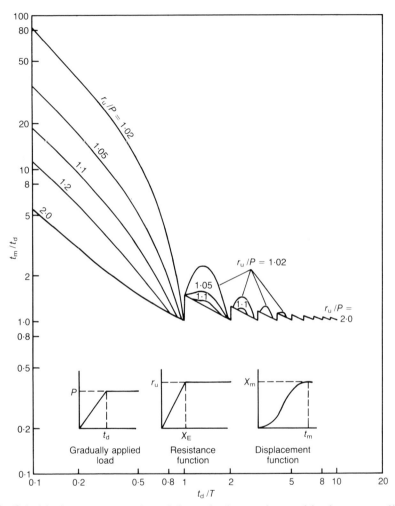

Fig C.4. Maximum response time of elasto-plastic, one degree of freedom system for gradually applied load (as per Fig. C.1.)

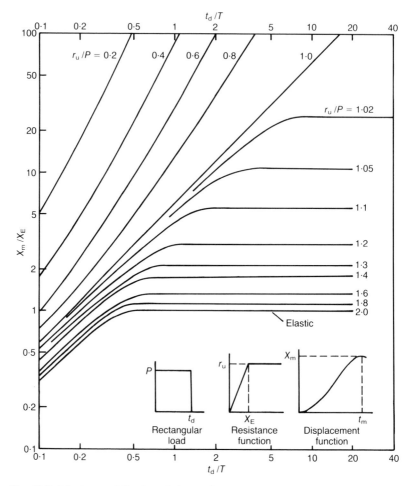

Fig. C.5. Maximum deflection of elasto-plastic, one degree of freedom system for rectangular load (as per Fig. C.1.)

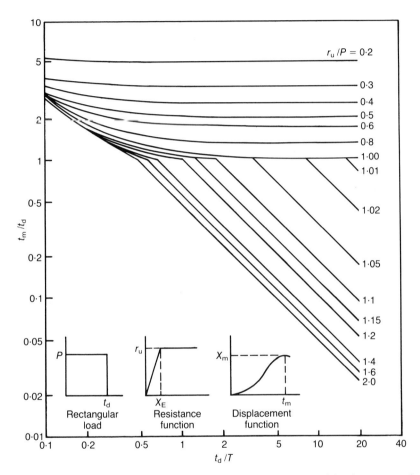

Fig. C.6. Maximum response time of elasto-plastic, one degree of freedom system for rectangular load (as per Fig. C.1.)

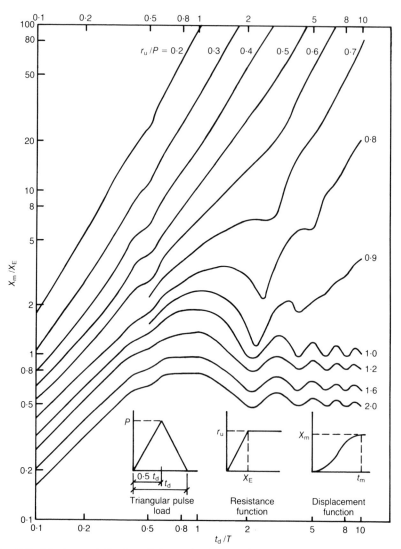

Fig. C.7. Maximum deflection of elasto-plastic, one degree of freedom system for triangular pulse load (as per Fig. C.1.)

APPENDIX C

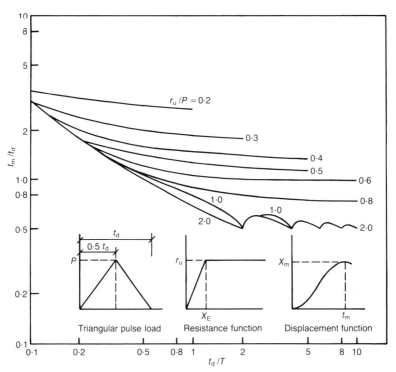

Fig. C.8. *Maximum response time of elasto-plastic, one degree of freedom system for triangular pulse load (as per Fig. C.1.)*

Appendix D. Design flow chart

Notes
1. As an approximate rule, reinforced concrete elements designed for protection category 1 ($\theta \leq 2°$) will exhibit a type 1 section and are likely to respond within the pressure–time (or dynamic) regime, whereas those designed for protection category 2 ($\theta > 2°$) will exhibit a type 2 section and are likely to respond within the impulsive regime. However, this is not a hard and fast rule and the response time of the element (t_m) must always be calculated and compared with the duration of the load (t_d) to check that the correct response regime has been assumed in design.
2. For the unusual case of protection category 2 and response within the pressure–time (or dynamic) regime, a type 2 section should be assumed at this stage.
3. For the unusual case of protection category 1 and response within the impulsive regime, a type 1 section should be assumed at this stage.
4. Structural steel sections usually respond within the pressure–time (or dynamic) regime. However, if the calculated response time (t_m) is large in relation to the duration of the applied load (t_d) an impulsive design procedure may be more appropriate.

BLAST EFFECTS ON BUILDINGS

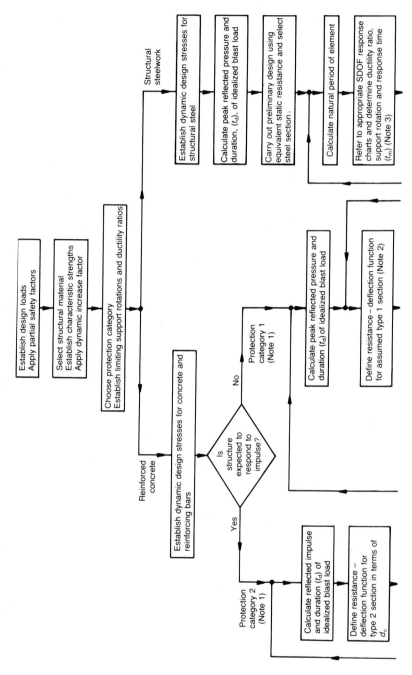

Fig. D.1 Flow chart of design procedures

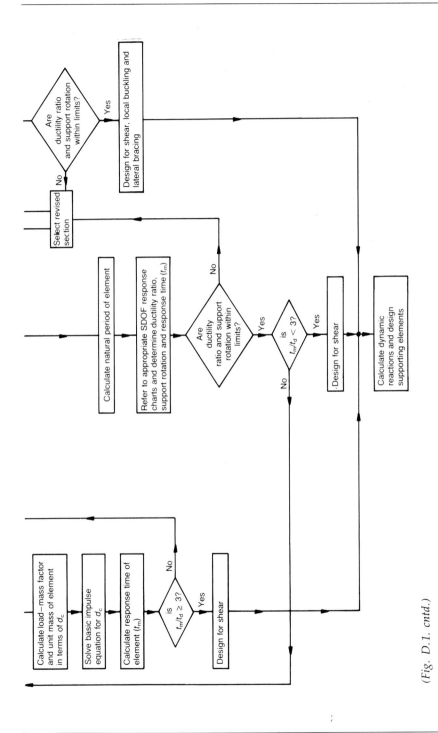

(Fig. D.1. cntd.)